Energy Technology 2010:

Conservation, Greenhouse Gas Reduction and Management, Alternative Energy Sources

Proceedings of symposia sponsored by the Light Metals Division of The Minerals, Metals & Materials Society (TMS)

Held during TMS 2010 Annual Meeting & Exhibition
Seattle, Washington, USA
February 14-18, 2010

Edited by:

Neale R. Neelameggham
Ramana G. Reddy
Cynthia K. Belt
Ann M. Hagni
Subodh Das

A Publication of

TMS

ENERGY TECHNOLOGY 2010

Related titles include:

- *Carbon Dioxide Reduction Metallurgy*
- *Energy Technology Perspectives: Conservation, Carbon Dioxide Reduction and Production from Alternative Sources*

HOW TO ORDER PUBLICATIONS

For a complete listing of TMS publications, contact TMS at (724) 776-9000 or (800) 759-4TMS or visit the TMS Knowledge Resource Center at http://knowledge.tms.org:

- Purchase publications conveniently online and download electronic publications instantly.
- View complete descriptions and tables of contents.
- Find award-winning landmark papers and webcasts.

MEMBER DISCOUNTS

TMS members receive a 30 percent discount on TMS publications. In addition, members receive a free subscription to the monthly technical journal *JOM* (both in print and online), free downloads from the Materials Technology@TMS digital resource center (www.materialstechnology.org), discounts on meeting registrations, and additional online resources to name a few of the benefits. To begin saving immediately on TMS publications, complete a membership application when placing your order or contact TMS:

Telephone: (724) 776-9000 / (800) 759-4TMS

E-mail: membership@tms.org or publications@tms.org

Web: www.tms.org

Energy Technology 2010:

Conservation, Greenhouse Gas Reduction and Management, Alternative Energy Sources

A Publication of **The Minerals, Metals & Materials Society (TMS)**
184 Thorn Hill Road
Warrendale, Pennsylvania 15086-7528
(724) 776-9000

Visit the TMS Web site at
http://www.tms.org

Printed in the United States of America
ISBN Number 978-0-87339-749-0

If you are interested in purchasing a copy of this book, or if you would like to receive the latest TMS Knowledge Resource Center catalog, please telephone (724) 776-9000, ext. 270, or (800) 759-4TMS.

TABLE OF CONTENTS
Energy Technology Perspectives 2010

Carbon Dioxide and Other Greenhouse Gas Reduction Metallurgy - 2010

TABLE OF CONTENTS
Energy Technology Perspectives 2010

Energy Conservation in Metals

TABLE OF CONTENTS
Energy Technology Perspectives 2010

Alternative Energy Resources for Metals and Materials Production Symposium

TABLE OF CONTENTS
Energy Technology Perspectives 2010

Carbon Management and Carbon Dioxide Reduction

Preface

Energy conservation has been found to be the essential first step with the highest payback in reducing greenhouse gases. Not only do you help the environment by reducing CO2 and other greenhouse gases, but you reduce industry costs. In today's economy, this is essential to keeping industry cost competitive.

The papers within the Energy Conservation symposium are an effort to share ideas to improve energy efficiency within our industry. Now in its third year, this year's papers look at energy conservation and energy efficiency from different ways. Conservation efforts can be found in improving both processes and equipment.

During 2008 and 2009, the symposium on minimizing carbon dioxide emissions by chemical reduction of oxides or physical minimization by other methods was called CO2 Reduction Metallurgy. The current symposium shown in these Proceedings is titled Greenhouse Gas Reduction & Management. The revised title fits the efforts going on in addressing the global warming issue from all greenhouse gases including carbon dioxide, water vapor, and other multi-atomic (more than two) gases which take part in the reflective radiation of heat from the atmosphere. In addition some papers address the carbon emission control which is based on economic trading means classified under the greenhouse gas management systems. The proceedings cover both aspects in several paper presentations from two sessions. The key-note presentation will discuss novel methods of converting greenhouse gases into solids by the use of electrical discharges (from non-carbon sources) which are similar to lightning type reactions occurring in nature.

A third new symposium has been added for the 2010 Energy Technology Session called Alternative Energy Resources for Metals & Materials Production. This symposium is geared to address the use of emerging alternate energy resources such as solar and wind power – besides the non-carbon based hydropower in Material Production processes. Some of the presentations planned for the symposium do not have manuscripts in these Proceedings. One of which is the keynote discussion on whether the use of non-fossil energy resources can help in address the global limitations on present day energy mineral resources. The attendees are encouraged to listen to presentations without manuscripts on exciting subjects like energy storage methodologies, use of solar thermal power for part of the process reaction energy in material manufacturing, etc.

We acknowledge the effort by the rest of the non-editor - co-organizers (Kotaru Ogura, Jiann-Yang 'Jim' Hwang, Xingbo Liu, Jean Pierre Birat, Brijendra Mishra, Aldo Steinfeld, R Palumbo, Mark Cooksey, Russell Hewertson, and Don Whipple) in collecting papers and presentations for the four sessions of the Energy Technology 2010.

This proceeding title delineates only the 4 sessions run by the energy committee of LMD & EPD and does not include other Energy Related symposiums during the Seattle Annual Meeting TMS – from other divisions – such as Nuclear energy symposiums, Clean Coal, Hydrogen Based Technologies, Fuel Cells and Materials for Energy Storage, Non-carbon materials in aluminum Production or Sustainable Technologies – each one of which is part of the term Energy Technology.

Editors:
Neale R. Neelameggham, Ramana G. Reddy, Cindy Belt, Ann Hagni and Subodh Das

Editors

Neale R. Neelameggham is the Technical Development Scientist for US Magnesium LLC. He has 37 years of expertise in magnesium production technology having been with the plant from its startup company NL Magnesium. Dr. Neelameggham's expertise includes all aspects of the magnesium process, from solar ponds through the cast house including solvent extraction, spray drying, molten salt chlorination, electrolytic cell and furnace designs, lithium ion battery chemicals and by-product chemical processing. In addition, he has an in-depth and detailed knowledge of alloy development as well as all competing technologies of magnesium production, both electrolytic and thermal processes worldwide. Dr. Neelameggham holds 13 patents and has several technical papers to his credit. As a member of TMS, AIChE, and a former member of American Ceramics Society he is well versed in energy engineering, biofuels and related processes. Dr. Neelameggham has served in the Magnesium Committee of LMD since its inception in 2000, chaired it in 2005, and has been a co-organizer of the magnesium Committee since 2004. In 2007 he was made a Permanent Co-organizer for magnesium Committee. He has been a member of Reactive Metals Committee, Recycling Committee and Programming Committee Representative of LMD. In 2008 LMD and EPD created the Energy Committee following the symposium on CO2 Reduction Metallurgy symposium initiated by him. Dr. Neelameggham was selected as the inaugural Chair for the Energy Committee with a two-year term. He is also a member of LMD council. Dr. Neelameggham's doctorate in extractive metallurgy from the University of Utah.

Ramana G. Reddy is an ACIPCO Endowed Chair Professor of Department of Metallurgical and Materials Engineering, The University of Alabama, Tuscaloosa, Alabama. His academic administration and research work experiences include: Department Head of the Department of Metallurgical and Materials Engineering and Associate Director of Center for Green Manufacturing at The University of Alabama; Professor and Chairman, Department of Chemical and Metallurgical Engineering, The University of Nevada, Reno; and Visiting Researcher at the Lawrence Berkeley Laboratory, University of California, Berkeley; Indian Institute of Technology, Bombay; and Argonne National Laboratory, Chicago. He has 28 years of teaching and research experience in the field of chemical and materials engineering. Professor Reddy has conducted projects involving thermodynamics and kinetics of metallurgical reactions; plasma processing of materials, molten salts and ionic liquids electrolysis, and fuel cells. He is the recipient of four three U.S. Patents and has published more than 325 research papers and 22 books including an undergraduate student textbook in thermodynamics. Professor Reddy has also delivered more than 200 invited lectures and research presentations in 23 nations. He is the associate editor of the Journal of Phase Equilibria and Diffusion and editor-in-chief of the International Journal for Manufacturing Sciences & Production. Professor Reddy has organized 24 national and international symposia including the National Science Foundation 2003 Design Service and Manufacturing Grantees and Research Conference. He was appointed the University of Alabama's coordinator for the National Space Science and Technology Center, and served as council member of the Alabama State Committee for the Department of Defense EPSCoR Programs. He is the recipient of many honors and awards, Professor Reddy received the 2008 Distinguished Member Award from the Society for Mining, Metallurgy & Exploration and the TMS Extraction & Processing Division 2002 Distinguished Lecturer Award. He is a Fellow of American Society for Metals International. He obtained his doctorate from the University of Utah.

Cynthia K. Belt is Senior Energy Engineer for Superior Industries. She manages the energy program to improve energy efficiency and reduce energy costs for the corporation. Cindy earned her Bachelor of Science, Mechanical Engineering from Ohio Northern University with additional graduate work in materials at Case Western Reserve and the University of Akron. She is Vice-Chair of the Energy Committee within TMS, a member of the Process Heating System Assessment Standard Committee for ASME, and a member of the energy committee of the American Foundry Society.

Ann M. Hagni has over twenty-five years of progressive, diversified applied mineralogy experience in management and materials science environments. Ann earned her BS in Geophysics, MS in Engineering Management, and Ph.D. in Geology, from the University of Missouri-Rolla (now Missouri University of Science & Technology). Dr. Hagni has successfully applied high tech research skills in failure analyses resolutions and development of new products and technologies, with over 25 authored publications and over 40 abstracted presentations. She has focused on the materials and energy industries of petroleum, coal mining, refractories, specialty materials (isotopically enriched boron for nuclear and semi-conductor industries), and construction (concrete, cement, aggregates, steel, drywall). Her characterization studies of raw materials and processed materials have emphasized improved metals recovery, hazardous waste containment, recycling of materials, depositional purity, ore deposition, and failure analyses. In addition, Ann has demonstrated accomplishments in streamlining complex technical systems into efficiency and improved productivity and profitability. Dr. Hagni is currently a Consulting Geoscientist as founder of Ann Hagni Consulting, LLC.

Dr. Subodh Das is the CEO & Founder of Phinix, LLC. Phinix, LLC, based in Lexington, Kentucky, USA, is dedicated to the promotion, development and implementation of low carbon footprint manufacturing technologies and carbon management and trading for the Minerals, Metals and Materials Industries to help Globally Responsible Resource Management. Dr. Das has over 30 years of global experience in manufacturing and technology areas covering wide disciplines including executive, project, operational, financial and technical management as well as being an accomplished scientist, engineer and inventor. Dr. Das is well recognized and respected expert and consultant to the aluminum industry specializing in the areas of technology, recycling and new product and process developments.

Energy Technology 2010:

Conservation, Greenhouse Gas Reduction and Management, Alternative Energy Sources

Carbon Dioxide and Other Greenhouse Gas Reduction Metallurgy - 2010

Organizers

Neale R. Neelameggham
Ramana Reddy
Jiann-Yang Hwang
Jean-Pierre Birat
Kotaro Ogura

Energy Technology 2010
Edited by: Neale R. Neelameggham, Ramana G. Reddy, Cynthia K. Belt, Ann M. Hagni and Subodh Das
TMS (The Minerals, Metals & Materials Society), 2010

OVERVIEW OF REMOVAL METHODS FOR CO_2 AND OTHER GREENHOUSE GAS AND DETAILS OF THE METHOD USING NON-THERMAL PLASMA

Marcela Morvová, Imrich Morva and Mário Janda

Comenius University, Faculty of Mathematics, Physics and Informatics, Mlynská dolina F2, 84248 Bratislava, Slovakia

Keywords: Non-thermal atmospheric plasma, Greenhouse gas mitigation, CO_2 removal, biomass and waste pyrolysis, hydrogen production

Abstract

The possibilities of greenhouse gas mitigation using various methods are brought out. The special case of utilization of mixture of CO_2 with other greenhouse gases in a synergetic way inside non-thermal plasma based system is described in detail.

Measurements on pilot system comprising pyrolysis chamber as alternative energy system (production of hydrogen and liquid tar) were made for various biomass and waste as the input material. The exhaust from this system was treated with non-thermal plasma gap suitable for up to 650Nm^3/hour of exhaust gas flow. Condensable portion of exhaust is liquefied in distillation unit; produced hydrogen is stored on-line inside nano sized carbon of our own production.

The CO_2 removal efficiency varies depending on exhaust composition between 45 to 98%. Other oxides as CO, NOx, SOx and several hydrocarbons and VOC are also removed. The main product connected with CO_2 removal is a solid powder of proteinoid nature. There exist strong probability, that precursor of the origin of life on the Earth from primary atmosphere were formed in the similar way as it is present in the electric discharge system used. Our system is equipped with on-line as well as ex-post analysis of all forms of products. We have also developed a system for recuperation of all forms of produced heat. It is stored inside as latent heat of waxes.

Introduction

Natural greenhouse effect on the Earth is caused mainly by water vapour with the average absolute concentration about 4%. Other gases able to absorb long-wave radiation and thus this causing greenhouse effect are CO_2, CH_4, N_2O, halocarbons and as precursors ozone and NO_X. Some of them we produce and introduce into atmosphere in large quantities during production of energy and in transport. This factor leads to permanent increase of average temperature of air, sea and oceans, so called global warming.

Nowadays large amount of data exists showing that the temperature of Earth in the last 150 years has been continually increasing. It represents about 03°C per ten years and the most of climatologist believe, that the increased concentration of greenhouse gases in the atmosphere is responsible for this effect.

The most important greenhouse gas is carbon dioxide CO_2. Majority of CO_2 emissions comes from electric power stations working with fossil fuels, cars and industry. Burning of fossil fuels contribute up to 80% to world wide anthropogenetic emissions of CO_2.

Overview of technological possibilities to mitigate greenhouse effect

CO_2 – besides water vapour – comprise majority of the gases causing greenhouse effect, is the major object gas for mitigation of greenhouse effect. In this process following steps are necessary:

1. Capture and separation of CO_2 from exhaust
   - chemical uptake into the solution,
   - membrane separation,
   - physical absorption into solid substances,
   - cryogenic separation – which requires combustion in pure oxygen instead of air as oxidant
2. Adjustment and transport to the place of final storage
3. Final storage

Final storage is possible to realize using various technical solutions into suitable geological formations. It includes:

1. Storage into deep sea (depth of deposit must be under 3000m),
2. Storage into deep saline reservoir,
3. Storage into depleted crude oil (or gas) deposits.

It is possible, to inject flue gas without CO_2 separation, to reach enhanced oil recovery from the action of anaerobic bacteria species in a period of about 4 years the storage of solid CO_2 is also considered as a possibility for storage inside carbon bed of the coal mines.

Natural and artificial methods of CO_2 utilization

Nature considers CO_2 as raw material suitable for building chemically interesting products, what we people use as foodstuff. Such production process is known as photosynthesis, eventually chemo- or thermo- synthesis according to the source of energy. Later when foodstuff spoils and can use it as fuel. It is necessary that also mankind gives helping hand to nature in this subject.

One of the interesting possibilities is to use biomass as fuel; it leads to change status of thermal process from producing CO_2 to circulating CO_2. As a fuel it is possible to use quickly growing terrestrial vegetation, aquatic vegetation (as plankton and algae) and waste. Additionally it is possible to use alternative fuels in transport. Here is necessary to think on alcohols for Otto engine, methyl esters of various oils for Diesel engine. For all forms of transport at near future interesting alternative are hydrogen based energetic system, first of all low temperature fuel cells (PEM).

Except for alternative energy sources it is still necessary to find appropriate ways for mitigation of CO_2 from existing sources. This is the place where our work can give an interesting contribution. A part of our work shows, that there are some correlations between processes we have studied (discharge treatment of combustion and pyrolysis gases) and processes taking place during the origin of life on Earth in primary atmosphere under influence of natural discharges.

Utilization of flue gas using spontaneously pulsing electric discharge, long period experimental results in various systems [4-10]

The behaviour of non-thermal atmospheric plasma (discharge system) in plasma chemical processes

Important properties of electric discharge systems working at pressures close to atmospheric are that the mean free path of particles is about 10 μm and the collision frequency is close to 10^{10} per second. This increases the probability of excitation due to repeated collision with electrons. The plasma-chemical reactions take place in several steps, which from the point of view of stepwise activation energy is very advantageous. The discharge initiates following activation factors: free electrons and ions which get ionized ionization by single or multiple electron impact, negative ions due to electron attachment on electronegative atoms or molecules, photons in VUV, UV and visible region of spectra, free radicals, excited atoms and molecules and metastables.

Described activated factors supply formation of activated complexes together with reacting (usually unwanted) compound. In the case of hydrocarbons and oxygen containing atmosphere partial oxidation always take place. In this process the second generation of free radicals (usually peroxy-acid radicals) is produced. The most important primary radical causing oxidative degradation and limiting the velocity of hydrocarbon decomposition process is the OH radical.

The type of chemistry applied in discharge is strongly dependent on the ratio between activation factors, "temperature" of involved species, present surfaces, residence time in discharge as well as the length of "after discharge" time i.e. the time between two pulses. When we take into account all these facts we have to consider various chemical reactions -- plasmochemical reaction with oxidation, reduction or both, polymerization, dry thermal decomposition (pyrolysis) and various more complicated mostly chain reactions.

In many cases heterogeneous processes on electrode surface as well as other surface on flowing potential play an important role.

Negative and positive ions generated in the primary processes are able to form clusters with water and other air impurities. On surface of such condensation centres the unwanted compound adsorb, then the heterogeneous reaction take place.

Usually the process is stopped only after polymerization or aggregation. After such processes the product leave the active volume like small aerosol particles.

Experimental systems – an overview of discharges

Non thermal non equilibrium plasma is realized it in numerous forms of discharges. Various modifications lead to relative bright scale of discharge equipment. Basic types of discharges burning at atmospheric pressure suitable for technical application are: arc, microwave, high frequency, laser induced, corona, dielectric barrier, ferroelectric packed bed and surface discharges, glid-arc, high pressure glow discharge & several forms of transient discharges.

In specific position are systems based on electron beam, which is used for various environmental applications up to industrial scale systems due to extremely high electron energy.

Transition discharges at atmospheric pressure

These types of discharges generate plasma laying on the boundary between non-thermal plasmas and thermal plasmas. Such plasmas can be generated in the transition regimes of streamer-induced discharges. These techniques often give interesting results comparable with typical non-thermal plasma techniques.

To this group of discharges count prevented spark, spontaneously pulsing corona to spark transition discharge, back corona, high pressure glow discharge HPGD (filamentary as well as diffuse forms) and glid arc and it is necessary to tell, and new forms are still developing. The development of these types of discharges depends on parameters of external circuit i.e. external resistance, capacity and inductance.

Prevented spark

Prevented spark developed by E. Marode belongs to the family of filamentary discharges induced by a streamer. The filamentary discharge structure is due to a tube of positive ions which results from the streamer trail along its way from the positive anode to the negative cathode. The ions keep all the electrons inside their tube and render a filamentary current distribution. As the streamer reaches the cathode, the conductive plasma bridges the electrodes and a spark may occur within this bridge. The spark intensity depends on the value of external impedances. All applications of non-thermal plasma for flue-gas cleaning are based on the ability of high pressure gas discharges to induce chemical activities within the gas volume and to convert the gaseous pollutants into somewhat non-hazardous. An attempt to promote a special type of the streamer-induced discharge called "prevented spark" has been suggested some years ago.

Spontaneously pulsing SPTD (corona to spark) transition discharge [4]

SPTD is a filamentary discharge induced by a streamer. The mechanism of its function is very similar to the prevented spark. At lower applied voltages we observe regularly vanishing streamers. As the applied voltage is higher, the small capacity *C* is more charged during the post-streamer relaxation phase. The energy stored in it is delivered to the discharge gap which forms a spark pulse with current maximum. The intensity of this pulse is given by this energy, which is not sufficient for the spark development. Once *C* is discharged which becomes very quickly, the high spark current falls to 0 due to the voltage fall on the external resistance *R*. For typical R~5 MΩ and C~30 pF the repetitive frequency of pulses f is in the order of several kHz, which causes a characteristic noise of the discharge. The discharge was supplied by a D.C. high voltage of both polarities.

This type of a streamer-induced discharge was first applied by M. Morvová et all. to CO_2 and CO depletion, nitrogen fixation, NO_X removal, VOC removal and combustion exhaust utilization producing SOLID AMORPHOUS PRODUCT.

Experimental system – gas cell discharge tube [4, 5, 10]

The discharge tubes used were of wire–plane configuration with various electrode materials. The stressed electrode was wire of 0.2 mm diameter wound on a glass tube. The non-stressed electrode was a 35 mm 100 mm plane. The inter-electrode distance was 11 mm. The discharge tube was situated in an IR absorption gas cell so that the electrodes were parallel to the optical axis of the gas cell, Fig.1. This gave the possibility of scanning *in situ* IR absorption spectra from the drift region of the corona discharge along the discharge tube. The IR reflection absorption spectra of the non-stressed plane electrode and IR absorption spectra of deposits on the stressed electrode could also be scanned after the completion of measurements and opening of the discharge tube.

Gas cell discharge system is possible to use so, that discharge application is reiterated with IR spectra measurement several times. This make possible to have better in sight into kinetic of plasma chemical reaction taking place inside system inclusive catalytic effects of electrodes.

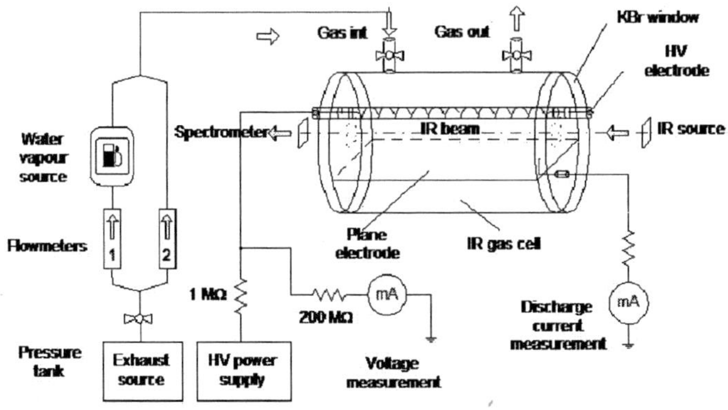

Fig. 1 Gas cell discharge system incl. sourcing and gas flow system

Experimental system – pilot system [6-9]

Complex of energetic- ecologic system was built in basic form in 1997 and it comprises of 3 basic modules:

- pyrolysis oven,
- gas flow system with on-line analysis of gas flow parameters and
- discharge system for exhaust cleaning.

The multifunctional discharge system used for the tests on pilot scale operates on gas flow volume of 50–650m^3/h and comprises in discharge system plurality of 24 discharge tubes connected parallel to each other. One discharge chamber consists of the copper rod with an internal thread (stressed electrode) and a coaxial cylinder (non-stressed electrode) with an inter-electrode distance of 6mm. The length of one discharge tube is 50cm.

High voltage source with dc high voltage of both polarities up to 20kV, maximum power 600W and maximum current 30mA was applied for discharge generation. The actual power depends on the composition of exhaust gases and actual gas flow.

Gradual improvements over the years have evolved the equipment with several online measurement capabilities as follows:

- system for isokinetic sampling for ex-post analysis,
- heat recuperator connected with tar distillation system giving possibility of liquid sampling,
- heat accumulation system using the melting [latent heat] of paraffin
- recuperation-distillation system for oils and paraffin
- system for fly carbon attachment
- system for fly solid discharge products attachment
- on-line hydrogen absorption system
- scrubbing and biochemical system for odor removal
- on-line chemical analysis system on 9 places of pilot system every 10 seconds for CO_2, CO, NO_X, CH_4, O_2, H_2 measurement using absolute measuring methods (thermal

conductivity for hydrogen, paramagnetic resonance for oxygen, other NDIR) in Maihak S700 analytic system. All on-line data are processed into PC.

For ex-post analysis we use complex of gas chromatograph Clarus 500, FTIR spectrometer Spectrum BX (both Perkin Elmer), mass spectrometer Nermag R10T and various optic and electron microscopes.

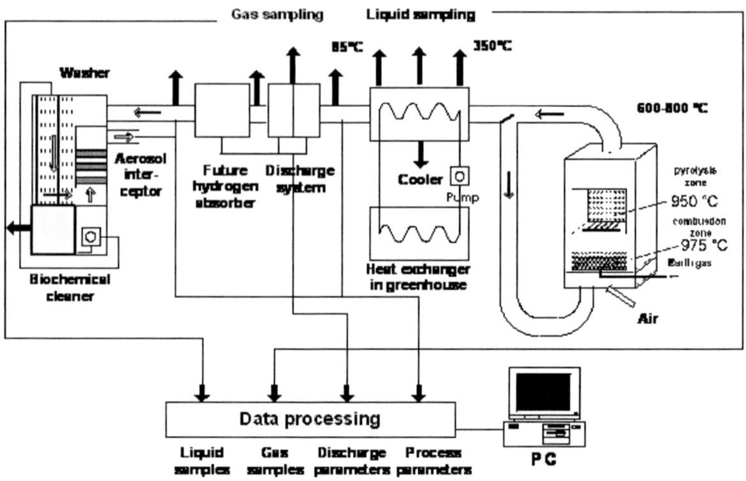

Fig. 2 Schema of pilot system

Measurements and results and discussion

In a gas cell discharge tube we have performed step by step measurements of the development with time of gas phase products and intermediates formed by an electric discharge in the introduced gas mixture. The infrared spectra from the area of the inter-electrode distance were scanned *in situ*. The studied gas mixture was introduced into the evacuated gas cell–discharge tube. The IR spectrum of the introduced gas was scanned. Then the discharge was applied for a defined time and the spectrum was scanned again. This procedure was repeated until no more changes in composition of the gas took place. Both positive and negative polarities of the discharge were applied.

During the time of discharge actions considerable changes of gas composition take place. So it was impossible to measure current–voltage characteristics. However, the actual values of the current and voltage were measured and processed in the PC. The applied voltage during the action of the glow corona was about 11 kV.

When the concentration of CO was high the discharge had a typical glow character. With an enhanced CO2 concentration the typical glow corona action alternated with a streamer-like spontaneously pulsing discharge with the fall of the voltage to about 1–2 kV.

After completion of gas phase measurements the non-stressed plate electrode was removed from the gas cell discharge tube and the reflection IR absorption spectrum of it was scanned. From this spectrum it was possible to identify the solid and liquid products on this electrode. The liquid products were additionally analyzed by the ATR technique Fig 5.

The time development of main product for oxygen rich atmosphere Fig 3 and oxygen pure atmosphere Fig. 4

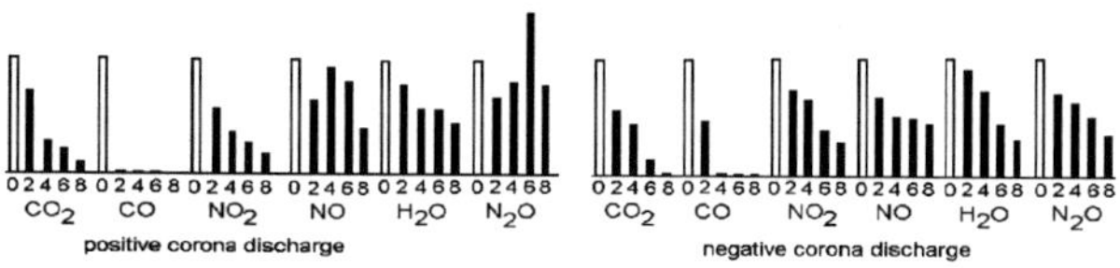

Fig. 3 Time development of main products for oxygen rich mixture of introduced gases and both polarities of applied discharge related to introduces concentration of each component of mixture (first not filled column represent introduced concentration)

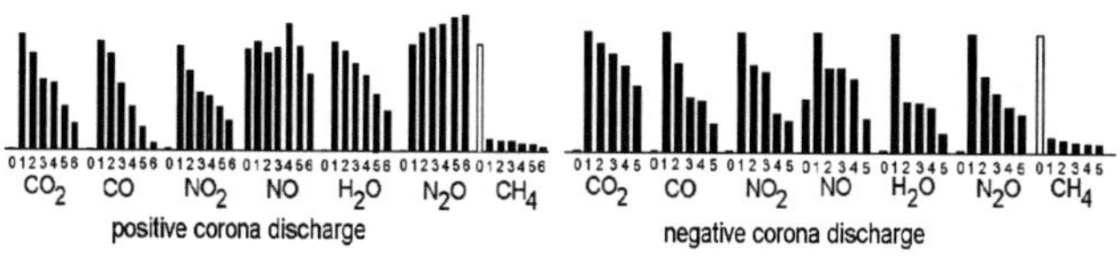

Fig. 4 Time development of main products for oxygen pure mixture (natural flue gases enriched with methane) of introduced gases and both polarities of applied discharge related to introduces concentration of each component of mixture (not filled column represent introduced concentration)

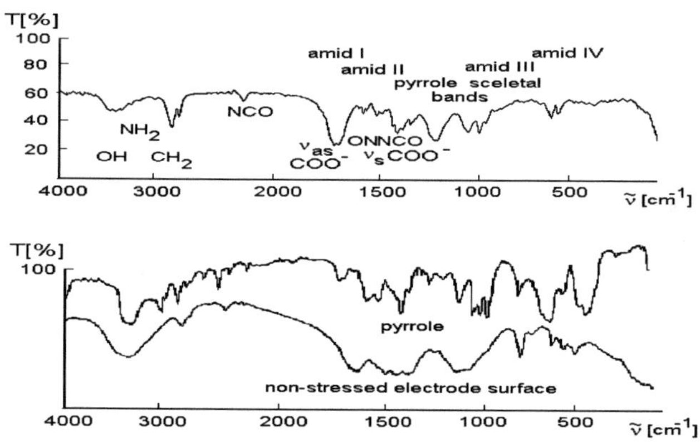

Fig 5 IR absorption spectra of solid product (spectrum up) separated from bulk of pilot system measured by KBr pellet technique in comparison to surface spectrum from brass electrode surface (spectrum down) from gas cell discharge tube measured by ATR technique.

IR spectra of bulk product confirm the formation of amino acids and heterocyclic unsaturated compounds due to presence of amide I, II, III bands (wave numbers 1700-1655, 1565-1400, 1300cm^{-1}).

The product is statistical polycondensate containing amino acids arginine, lysine, histidine, methionine, glycine alanine, serine, aspartic acid. This information was gained by comparing IR spectra of product with calibration IR absorption spectra of pure components [12, 13] and also from HPLC and thermo gravimetric analysis.

The organo metallic surface product has important catalytic properties particularly due to enhanced dielectric constant. From ATR spectra is seen the presence of oxamidato complexes with known ferroelectric properties and oligo pyrrole type of compounds with probable catalytic activity. It is known that linear and cyclic tetrapyrrole compounds are important part of photosynthetic chromophores (linear phycocynine, cyclic chlorophyll). Comparison of IR absorption spectrum of non stressed electrode surface and pyrrole can be seen from Fig 5.

The combination of gas cell measurements with pilot plant measurements gives us possibility to have better inside into plasma chemical processes taking part inside discharge system.

The involved chemistry during final product formation was extracted from step by step kinetic studies made inside gas cell discharge tube. We have divided the process into several important steps:

1. activation,
2. formation of energy rich intermediate species,
3. formation of catalytic spots on electrode surfaces,
4. volume reactions in non equilibrium plasma condition,
5. surface reactions on electrodes
6. solid final products

The most important activation step is excitation of molecular nitrogen by repeated electron impact in electric discharge using in total energy close to 6.5eV.

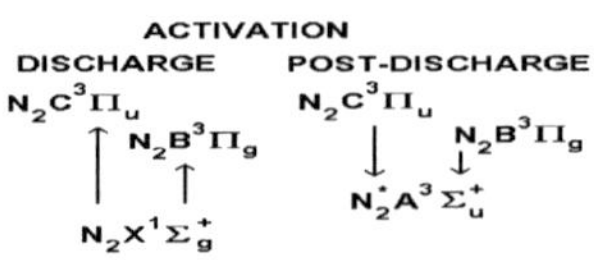

$$N_2 \quad X^1\Sigma^+_g \quad \rightarrow \quad N_2 \quad A^3\Sigma^+_u$$

The described electronic state $N_2 \quad A^3\Sigma^+_u$ has a life time between 1.3-1.9 sec [11] and that is what can be involved with a great probability into chemical reactions development. The N_2 activation step is followed by its incorporation into CO_2 connected with the formation of two most important active radicals namely ON-NCO and -NCO.

$$O = C = O + N_2 \ A^3\Sigma^+_u \longrightarrow O{=}N{-}N{=}C{=}O$$

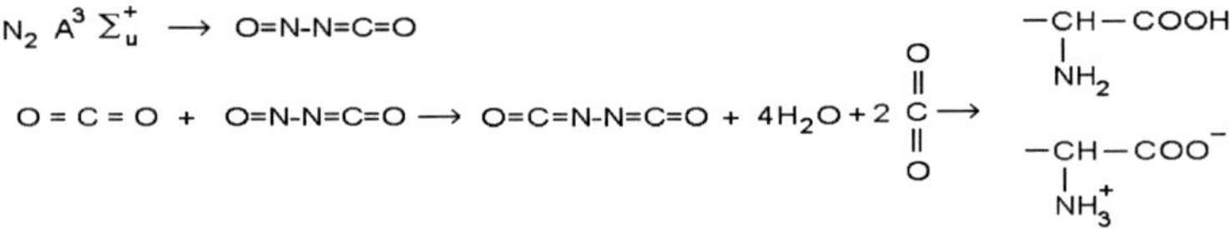

The development of all further products is based on the reaction with of NCO (linear amino acids) and ON-NCO (heterocycle containing amino acids) radicals.

The second important activation step is H_2O dissociation and/or dissociative ionization on the ferroelectric spots from oxamidato complexes on electrode.

$$H_2O \rightarrow H + OH \text{ or } H_2O \rightarrow H^+ + OH^-$$

By reaction of -NCO and ON-NCO radicals with water dissociation products various amino acids and urethanes are formed. Due to influence of electric field CO_2 is deformed into carboxylate ion. By the reaction of carboxylate ion with water cyclic dimer of carboxylic acid is formed and finally dicarboxylic acids (malonic, succinic) and hydroxy acids (apple, citric and tartaric). The urethanes create polyamides, in the case of trimerization barbituric acid was found in products. Overview of possible product formation is on Fig 6.

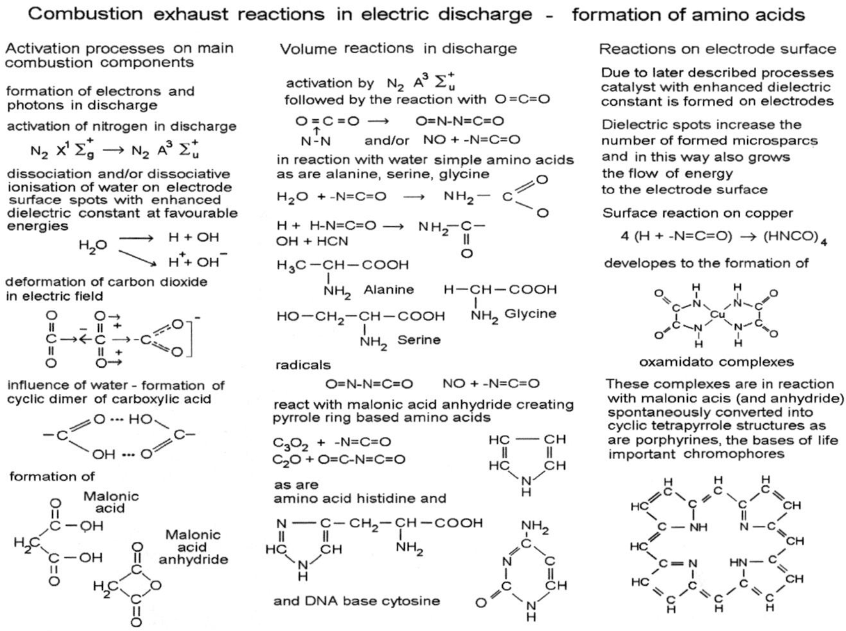

Fig 6 The overview of product based on amino acid groups formed inside electric discharge with a flue gas – pyrolysis gas – primitive atmosphere composition

Amino acids can also be formed on both electrodes. The quenching of active species take place in electric field inside drift zone of discharge and leads to the formation of clusters and finally polycondenzation of amino acids together with other products into amorphous amino acid condensate. The final product of process is a powder with fractal structure on microscopic level, with low specific weight, and is water insoluble. The main component (95%) of solid product is amorphous condensate of amino acids with about 5% of metal organic compounds with catalytic properties. Microscopic photograph of this product is in Fig 7, 8.

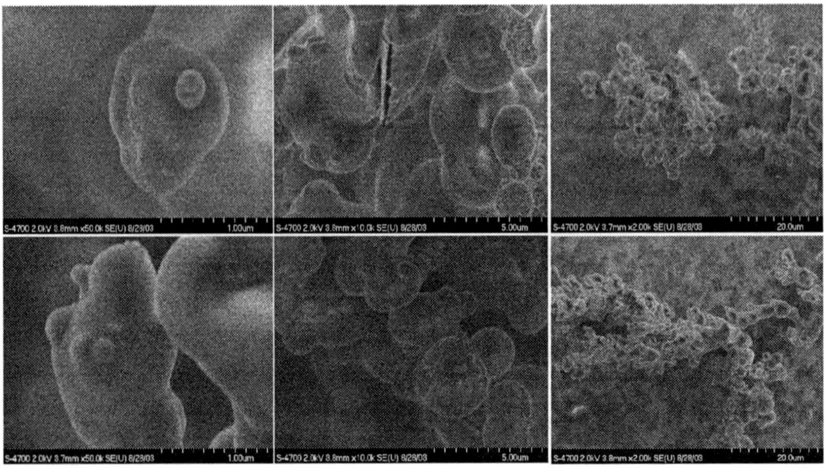

Fig 7 Micelle like proteinoid products formed due to action of spontaneously pulsing electric transition discharge on flue gases (sample from pilot system without pyrolysis process)

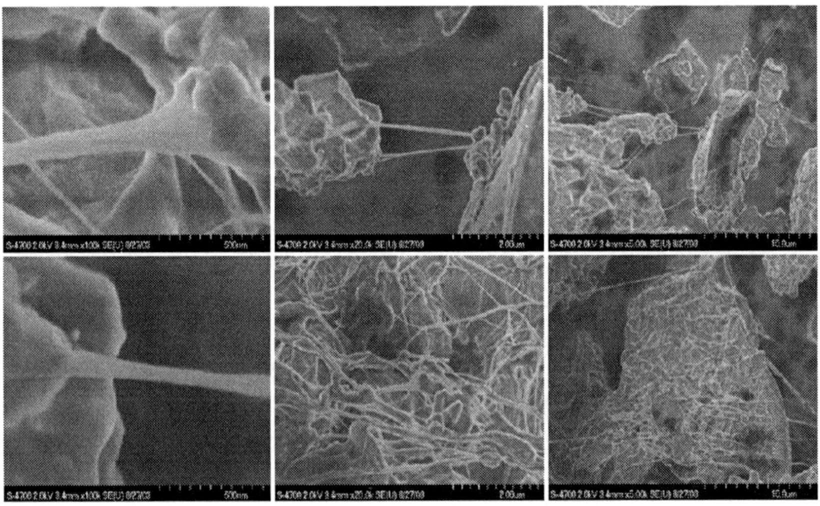

Fig 8 Micelle like proteinoid products formed due to action of spontaneously pulsing electric transition discharge on pyrolysis gases (sample from pilot system with pyrolysis process)

Pyrolysis process has very dynamic character. The changes inside pyrolysis chamber are very quick and lead to rapid changes in gas concentration for example in one second the gas flow changes from 150 to 650 m^3/hour due to bubble like character of process. This factor causes the need to use not only isokinetic sampling but also to scan the whole spectra in one minute interval before and after discharge system. As far as the isokinetic introducing of exhaust into cell need 5 minutes each cell, scanning of spectra 4 minutes each and full pyrolysis process take 29 minutes it was necessary to use 20 pyrolysis cycles to have such spectra development in one minute interval before and after discharge. Gained spectra are on Fig 9. The CO_2 concentration calculated from such spectra are on Fig 10. The difference between black and red line show the actual efficiency of the discharge on the carbon dioxide removal.

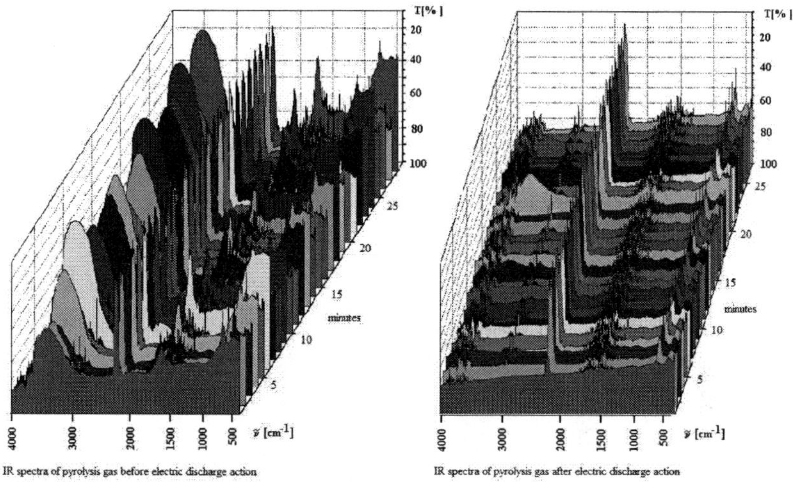

Fig 9 IR absorption spectra of pyrolysis gases before and after discharge system in minute intervals

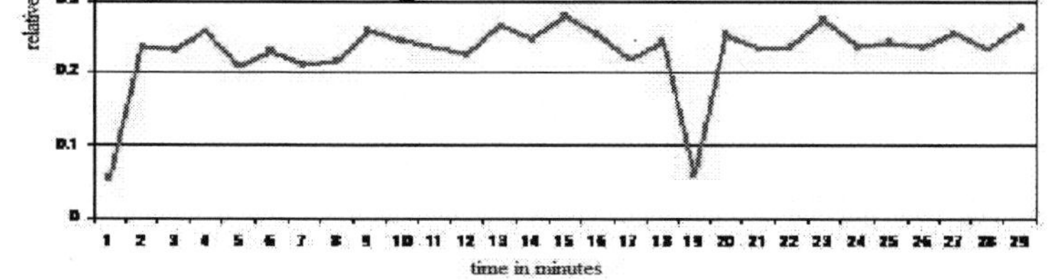

Fig 10 The CO_2 concentration development in pyrolysis gases before and after discharge.

Conclusions

We have developed method and equipment for continuous solidification of CO_2 and other exhaust components using the synergetic effect of electric discharge and catalysis induced by metal organic product formed on copper and or brass electrodes. The main advantage of the method is that all the involved processes taking place in single equipment, with a lower investment and operation cost necessary for complete cleaning of exhaust.

The efficiency of CO_2 solidification depends on the composition of introduced exhaust but range from 40-100%. Side effect efficiency on removal of NO_X and CH_X can reach 95%. Energetic cost of process is very low about 2,3-4,7 Wh/m^3, i.e. 8,3-16,9 kJ/m^3.

Product is a powder, water insoluble proteinoid, able to convert into jell by long lasting hydrolysis or by enzymatic way. As far as described processes are close connected with the origin of life on Earth, it is probable that it will be possible to implement such process in large extent, what is necessary for mitigation of greenhouse effect and global warming.

Acknowledgement

The authors wish to thank for support of Slovak Research and Development Agency APVV, project No. 0267- 06.

References

[1] Miller, S.L, Science 117 (1953), p. 528-529

[2] L.E.Orgel: *Orig. Life Evol. Biosphere* 28 (1998), pp. 227-234

[3] A.Pappelis & S.Fox at http://www.siu.edu/~protocell/photos.htm

[4] M. Morvová: J.PhysD: Appl.Phys 31 (1998), pp. 1865-1874

[5] M. Morvová: Czech.J.Phys. Vol.49, No.12 (1999), pp. 1703-1720

[6] M. Morvová, I. Morva, F. Hanic: Greenhouse Gas Control Technologies, Proceedings of the 4th International Conference on Greenhouse Gas Control Technologies, Ed. B.Elliasson, P. Riemer, A. Wokaun, Elsevier Science, Oxford, UK, 1999, ISBN: 008-043018-X, pp. 137-142

[7] M. Morvová, I. Morva, F. Hanic: Plasma Processes and Polymers pp. 403-412, Ed. R. d Agostino, P. Favia, C. Oehr, M.R. Wertheimer, ISDN 3-527-40487-2, 2005 Wiley-VCh Verlag GmbH

[8] M. Morvová, F. Hanic, I. Morva, J. Thermal Anal. Calorimetry 61 (2000) 273

[9] F. Hanic, M. Morvová, I. Morva, J. Thermal Anal. Calorimetry 60 (2000) 1111.

[10] Janda M., Machala Z., Morvová M., Morva I.: Study of Plasma Induced Chemistry by DC Discharges Above Water Surface in CO2/N2/H2O Mixture, Origins of Life and Evolution of Biospheres, Vol.38, No.1 (2008), pp. 23-35

[11] Lofthus, P.H. Krupenie, J. Phys. Chem. Ref. Data 6 (1977) 113.

[12] DMS, *Working atlas of infrared spectroscopy*, Butterworths&Co (Publishers) Ltd., London, 1972, ISBN 0 407 69999 6

[13] Holly S., Sohar P., *Absorption spectra in the infrared region*, Theoretical and technical introduction, Ed. by L.Lang and W.H.Prichard, Academiai Kiado Budepest, 1975, ISBN 963 05 01872

Energy Technology 2010
Edited by: Neale R. Neelameggham, Ramana G. Reddy, Cynthia K. Belt, Ann M. Hagni and Subodh Das
TMS (The Minerals, Metals & Materials Society), 2010

Synthetic Fuel Production Utilizing CO_2 Recycling as an Alternative to Sequestration

Joseph Hartvigsen[1], S. Elangovan[1], Lyman Frost[1], Carl Stoots[2], James O'Brien[2], J. S. Herring[2], Manohar Sohal[2], Grant Hawkes[2]

[1]Ceramatec, Inc. 2425 South 900 West, Salt Lake City, UT 84119-1517, USA
[2]Idaho National Laboratory, PO Box 1625, Idaho Falls, ID 83415-3860, USA

Keywords: carbon dioxide, high temperature electrolysis, Fischer-Tropsch synthesis, greenhouse gas reduction, climate change

Abstract

Pressing environmental and energy security concerns have brought much needed attention to economic production of hydrogen as the secondary energy carrier for non-electrical markets as well as to meet increasing demand for crude upgrading and desulfurization. While steam reforming of natural gas is currently the dominant method of hydrogen production, it is depleting a valuable fossil fuel while emitting green house gases. Thus, in the long run, efficient, environmentally friendly, and economic means of hydrogen production using nuclear and renewable energy needs to be developed. High temperature electrolysis using solid oxide electrolyte cells is an exceptionally efficient means of generation of high purity hydrogen. It is also capable of directly converting steam and CO_2 into synthesis gas. This co-electrolysis process enables conversion of CO_2 recovered from concentrated industrial sources into a feedstock for synthetic fuels production. Use of CO_2 provides a means of storing hydrogen in a concentrated and convenient form – a hydrocarbon. This is also an effective storage medium for intermittent renewable energy sources such as wind and solar.

Ceramatec, Inc and the Idaho National Laboratory are applying solid oxide fuel cell technology to the production of synthesis gas by high temperature co-electrolysis of steam and carbon dioxide. This technology utilizes solid oxide fuel cell stacks to electrochemically remove oxygen from steam and CO_2, leaving a clean stream of hydrogen and carbon monoxide suitable for fuel synthesis. By varying the relative concentrations of carbon dioxide and steam, the hydrogen to carbon monoxide product ratio can be controlled as required by the desired type of synthetic fuel being produced from the syngas. The resultant synthesis gas has been used as the feedstock to produce synthetic methane and Fischer Tropsch liquid fuels.

Alkaline and proton exchange membrane electrolysis cells typically operate below 100°C, while solid oxide electrolysis cells operates in the 750-850°C range. This higher temperature improves the efficiency significantly. Practically, it translates into a lower cell operating voltage, 1.1-1.4V for high temperature electrolysis vs. ~1.8V for water electrolysis. Hydrogen production is stoichiometric with current in all types of electrolyzers.

Carbon dioxide is a valuable resource as a feedstock for energy storage as a hydrocarbon. There is a vast amount of renewable energy that cannot be effectively utilized as grid electricity, but which could be used to convert CO_2 to hydrocarbons, offering a solution to the even more difficult (than supplying grid power) challenge of production of sustainable, domestic transportation fuels. Conventional electrolysis and chemical processes could be deployed to accomplish this conversion. However, high temperature co-electrolysis of CO_2 and steam to produce synthesis gas using solid oxide fuel cell technology has the potential of improving the efficiency by nearly 50%. A domestic synfuel industry will enable a much larger reliance on intermittent renewable energy than can be accommodated by conventional electric demand profiles while increasing our energy security.

Introduction

Based on ice core data and atmospheric measurements of carbon dioxide content, it has been reported that there have been significant increases in the amount of carbon dioxide since pre-industrial time[i]. Measurements indicate a steady rise of about .9 ppmv (parts per million by volume) per year from a pre-industrial value of ~280 ppmv to a current value above 380 ppmv (NOAA, July, 2009). Several scientists (University California San Diego, et al) have hypothesized that the anthropomorphic CO_2 generated from various industrial sources can be shown to correlate well with the current increase in temperature associated with climate change. Data gathered and reported by the Intergovernmental Panel on Climate Change in their report filed in November, 2007, indicates that fossil fuel use accounts for over 50% of total anthropogenic greenhouse gas emissions in 2004 (see page 36 of referenced report).

As an outgrowth of work started under the U.S. DOE's nuclear hydrogen initiative (NHI), Ceramatec, Inc and the Idaho National Laboratory (INL) have been developing high temperature co-electrolysis of mixtures of steam and CO_2 to produce syngas.

$$H_2O + CO_2 \longrightarrow H_2 + CO + O_2, \quad (1)$$

This synthesis gas has been used to produce substitute natural gas (SNG) and Fischer-Tropsch liquid fuels in the laboratory.

Early results of this work were presented at the TMS 2008 CO_2 Reduction Metallurgy Symposium in New Orleans.[ii] This paper reports additional results, however some explanations are repeated for the reader's convenience and to maintain the flow of thought. Over the past 6 years the scale of testing has increased from button cells to short stacks, to multi-stack modules and most recently to a multi-module unit incorporated into an integrated process skid as shown in Figure 1. This integrated laboratory scale

(ILS) demonstration moved beyond testing of just the electrochemical stack, and proved many of the key engineering elements that will be required in a practical industrial implementation of this technology. An independent review team assembled by the DOE to rank technologies being developed by the Nuclear Hydrogen Initiative (NHI), recommended high temperature electrolysis for continued development well ahead of the competing thermochemical and hybrid thermochemical process. Even though hydrogen production was the primary objective of the NHI program, one of several determining factors noted by the independent review team was the applicability of high temperature electrolysis to conversion of carbon dioxide to synthetic fuels.

Button cell, 3.2 cm^2 (2003) | 10-cell stack, 640 cm^2 (2004) | 120-cell half-module, 7,680 cm^2 (2006)

Integrated Laboratory Scale 720 cells, 3 modules, 46,080 cm^2, (2008)

ID	Component
1	Hot zone enclosure lid
2	Power supply and instrument racks
3	Electrical distribution cabinet
4	Data acquisition and control monitors
5	Deionized water system
6	Steam generator
7	Steam and H_2 superheaters
8	Air compressor
9	Patch panel
10	Product finned cooler
11	Steam condenser
12	Mass flow controllers
13	H_2 vent
14	Air and O_2 vent

Figure 1. Progression of High Temperature Electrolysis Testing at Ceramatec and the INL

To further demonstrate integration of electrolysis and fuel synthesis processes, Ceramatec is in the final stages of commissioning a syngas compression and storage system (Figure 2). This will complete the link from carbon dioxide to synthetic liquid fuels by supplying Ceramatec's modular Fischer-Tropsch reactor (Figure 3) with pressurized syngas produced by co-electrolysis of steam and carbon dioxide. While Ceramatec has previously demonstrated production of syngas by co-electrolysis, and methanation of that syngas at atmospheric pressure, production of liquid synfuels has relied on bottled syngas. The INL is in the process of implementing a similar system.

Figure 2. Compression and storage of co-electrolysis products for fuel synthesis at Ceramatec

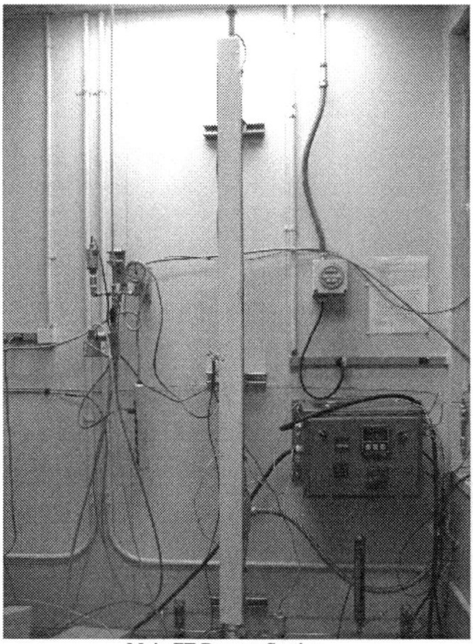

Main FT Reactor Section

Synthetic Fuel From CO_2 Recycling

Product Collection & Fractionation

Figure 3. 1.8m Modular Fixed Bed Fischer-Tropsch Reactor

Bio-derived synfuels are a current focus among policy makers as they have the potential to supplement petroleum fuels without replacing the existing vehicle fleet and refueling infrastructure. Interest has shifted away from early contenders such as corn-ethanol as more is understood about their overall economics and efficiency shortcomings. This has led to increased interest in cellulosic biomass processing to fuels. Both biological and chemical processing routes are being developed for cellulosic biomass conversion. The gasification route converts cellulosic biomass; such as barley and wheat straw, corn stover, or wood chips; into synthesis gas and ultimately into a liquid synthetic fuel for the transportation sector (biomass to liquids, or BTL). However similar to CTL (coal to liquids), nearly two-thirds of the feedstock carbon is lost as CO_2 in a conventional BTL process.

Co-electrolysis coupled with biomass gasification provides additional synthesis gas production enabling conversion of virtually all of the carbon content of the feedstock into synthetic fuels. This process has been termed Bio-Syntrolysis by investigators working on this technology at the INL (Figure 4). In Bio-Syntrolysis, biomass is first gasified to produce syngas directly. Gasification also provides a heat source and the carbon source (i.e. captured carbon dioxide) for additional syngas production via High Temperature Co-Electrolysis (HTCE). This syngas production is in addition to the syngas produced directly by biomass gasification, and requires an external input of renewable or nuclear electrical power. Any non-fossil source of electrical power such as wind, solar (i.e. photovoltaic [PV] or concentrator photovoltaic [CPV]) or hydro is suitable to power the HTCE step and achieve fuel production from a process completely free of carbon emissions. A byproduct of HTCE is hot oxygen, which can be used in place of air by the gasifier, thereby facilitating CO_2 recovery from the gasifier.

Using renewable generated electric power to convert the CO_2 byproduct of biomass gasification into synthesis gas and then to liquid fuel will increase the yield of biomass derived liquid fuels by a factor of 2-3. This becomes an effective means of storing energy from intermittent renewable energy sources such as wind and solar, in which case renewable energy comprises a majority of the energy embodied in such a bio-derived fuel. One problem facing both wind and biomass developers is that of transmission and transportation. These are both rather thinly distributed energy resources that are typically located far from the major urban areas having the highest demand for the resource. Bio-syntrolysis has the potential to couple these resources to produce a storable and transportable energy.

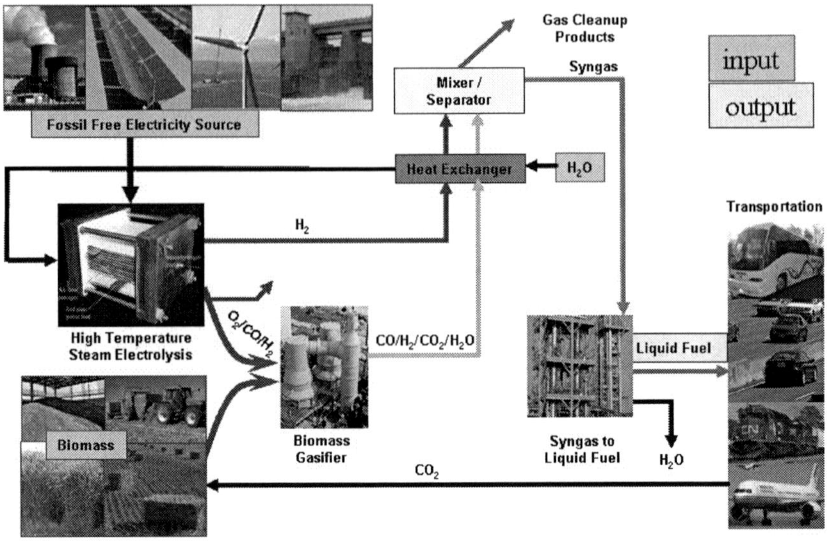

Figure 4. Bio-Syntrolysis Energy Process Scheme

High Temperature Electrolysis Fundamentals

A thermodynamic analysis of high temperature electrolysis begins by considering a control volume surrounding an isothermal electrolysis process, as shown in Figure 5. As illustrated in the figure, both heat and work interactions cross the control volume boundary.

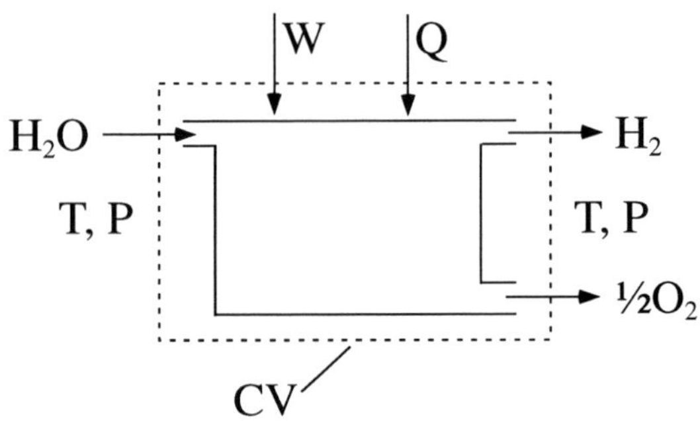

Figure 5. Thermodynamic control volume about an electrolysis process

The first law for this process is given by:

$$Q - W = \Delta H_R \tag{2}$$

For reversible operation,

$$Q_{rev} = T\Delta S_R \tag{3}$$

such that

$$W_{rev} = \Delta H_R - T\Delta S_R = \Delta G_R \tag{4}$$

The thermodynamic properties appearing in Eqn. (4) are plotted in Fig. 6 as a function of temperature for the H_2-H_2O system from 0°C to 1000°C at standard pressure. This figure is often cited as a motivation for high-temperature electrolysis versus low-temperature electrolysis. It shows that the Gibbs free energy change, ΔG_R, for the reacting system decreases with increasing temperature, while the product of temperature and the entropy change, $T\Delta S_R$, increases. Therefore, for reversible operation, the electrical work requirement decreases with temperature, and a larger fraction of the total energy required for electrolysis, ΔH_R, can be supplied in the form of heat, represented by $T\Delta S_R$. Since heat-engine-based electrical work is limited to a production thermal efficiency of 50% or less, decreasing the work requirement results in higher overall thermal-to-hydrogen production efficiencies. Note that the total energy requirement, ΔH_R, increases only slightly with temperature, and is very close in magnitude to the lower heating value of hydrogen. The ratio of ΔG_R to ΔH_R is about 93% at 100°C, decreasing to only about 70% at 1000°C. Operation of the electrolyzer at high temperature is also desirable from the standpoint of reaction kinetics and electrolyte conductivity, both of which improve dramatically at higher operating temperatures. Potential disadvantages of high-temperature operation include the limited availability of very high temperature process heat and materials issues such as corrosion and degradation.

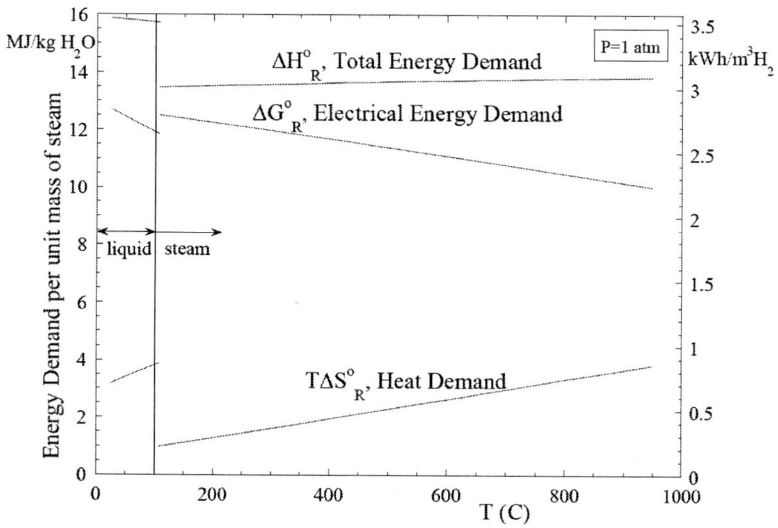

Figure 6. Standard-state ideal energy requirements for electrolysis as a function of temperature.

The solid-oxide electrolysis cell is a solid-state electrochemical device consisting of an oxygen-ion-conducting electrolyte (e.g., yttria- or scandia-stabilized zirconia) with porous electrically conducting electrodes deposited on either side of the electrolyte. A cross-section of a planar design is shown in Fig. 7. The design depicted in the figure shows an electrolyte-supported cell with a nickel cermet cathode and a perovskite anode such as strontium-doped lanthanum manganite (LSM). In an electrolyte-supported cell, the electrolyte layer is thicker than either of the electrodes. The flow fields conduct electrical current through the stack and provide flow passages for the process gas streams. The separator plate or bipolar plate separates the process gas streams. It must also be electrically conducting and is usually metallic, such as a ferritic stainless steel.

As shown in the figure, a mixture of steam and hydrogen at 750-950°C is supplied to the cathode side of the electrolyte. Note that cathode and anode sides are opposite to their fuel-cell-mode names. For readers familiar with SOFC materials, electrochemical convention dictates that the SOFC anode becomes the SOEC cathode at the instant of current reversal. This often causes confusion when discussing "the anode" or "the cathode", so frequently the terms hydrogen electrode and oxygen electrode are used since they are invariant with current direction. The half-cell electrochemical reactions occur at the triple-phase boundary near the electrode/electrolyte interface, as shown in the figure. Oxygen ions are drawn through the electrolyte by an applied electrochemical potential. The ions liberate their electrons and recombine to form molecular O_2 on the anode side. The inlet steam-hydrogen mixture composition may be as much as 90% steam, with the remainder hydrogen. Hydrogen is included in the inlet stream in order to maintain reducing conditions at the cathode. The exiting mixture may be as much as 90% H_2. It is not desirable to attempt to operate the stack with a higher outlet composition of hydrogen because there is a risk of local steam starvation which can lead to significantly reduced cell performance. Furthermore, based on detailed analyses of large-scale HTE systems, the overall hydrogen production efficiency is almost flat above about 50% steam utilization[iii]. Product hydrogen and residual steam is passed through a condenser or membrane separator to purify the hydrogen.

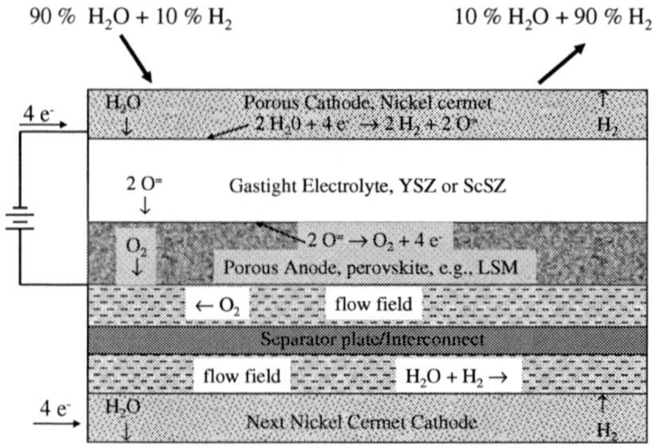

Figure 7. Schematic of a high temperature electrolysis stack.

In order to accomplish electrolysis, a voltage must be applied across the cell that is greater in magnitude than the open-cell potential. The standard-state open-cell potential is given by:

$$V^o = \frac{\Delta G_R^o}{jF} \tag{5}$$

where j is the number of electrons transferred per molecule of hydrogen produced. For the steam-hydrogen system, in which the $O^=$ ions are transported through the solid-oxide electrolyte, $j = 2$. The standard-state open-cell potential applies to the case in which pure reactants and products are separated and at one standard atmosphere pressure. In most practical HTE systems, the incoming steam is mixed with some hydrogen and possibly some inert gas. Residual steam can be removed from the product by condensation. On the oxygen-evolution side of the cells, air is often used as a sweep gas, so the oxygen partial pressure is only about 21% of the operating pressure. In addition, the electrolysis system can operate at elevated pressure. In order to account for the range of gas compositions and pressures that occur in a real system, the open-cell (or Nernst) potential can be obtained from the Nernst equation, which can be written as:

$$V_N = V^o - \frac{R_u T}{jF} \ln\left[\left(\frac{y_{H2O}}{y_{H2} y_{O2}^{1/2}}\right)\left(\frac{P}{P_{std}}\right)^{-1/2}\right] \tag{6}$$

Operation of a solid-oxide stack in the electrolysis mode is fundamentally different than operation in the fuel-cell mode for several reasons, aside from the obvious change in direction of the electrochemical reaction. From the standpoint of heat transfer, operation in the fuel-cell mode typically necessitates the use of significant excess air flow in order to prevent overheating of the stack. The potential for overheating arises from two sources: (1) the exothermic nature of the hydrogen oxidation reaction, and (2) ohmic heating associated with the electrolyte ionic resistance and other loss mechanisms.

Conversely, in the electrolysis mode, the steam reduction reaction is endothermic. Therefore, depending on the operating voltage, the net heat generation in the stack may be negative, zero, or positive. This phenomenon is illustrated in Figure 8. The figure shows the respective internal heat sink/source fluxes in a planar solid-oxide stack associated with the electrochemical reaction and the ohmic heating. The ohmic heat flux (W/cm^2) is given by:

$$q''_{Ohm} = i^2 ASR = i(V_{op} - V_N) \tag{7}$$

where i is the current density (A/cm^2), V_N is the mean Nernst potential for the operating cell, and ASR (Ohm·cm^2) is the area-specific resistance of the cell. The reaction heat flux is given by:

$$q''_R = \frac{i}{2F}(T\Delta S_e) = \frac{i}{2F}(\Delta G_e - \Delta H_R) \tag{8}$$

where ΔS_e is the entropy change for the actual electrolysis process, accounting for the reactant and product partial pressures.

The net heat flux is also shown in Fig. 8. An area-specific resistance of 1.25, an operating temperature of 1200 K, and hydrogen mole fractions of 0.1 and 0.95 at the inlet and outlet, respectively, were assumed for these calculations. In the fuel-cell mode, the net heat flux is always positive and increases rapidly with operating voltage and current density. In the electrolysis mode, the net heat flux is negative for low operating voltages, increases to zero at the "thermal-neutral" voltage, and is positive at higher voltages and current densities.

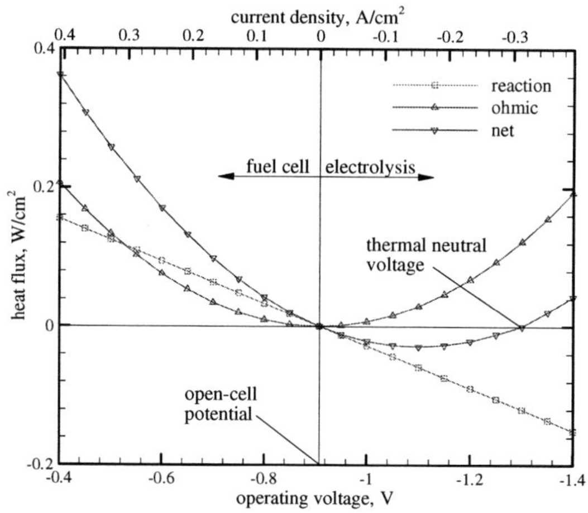

Figure 8. Comparison of thermal contributions in fuel cell and electrolysis modes.

The thermal-neutral voltage can be predicted from direct application of the rate-based First Law to the isothermal system shown in Figure 5:

$$\dot{Q} - \dot{W} = \Delta \dot{N}_{H_2} \Delta H_R \tag{9}$$

where, from Faraday's law,

$$\Delta \dot{N}_{H_2} = I/2F \tag{10}$$

Letting $\dot{Q} = 0$ (no external heat transfer), $\dot{W} = IV_{tn}$, yielding:

$$V_{tn} = \Delta H_R /2F \tag{11}$$

Note that the reaction heat flux of Eqn. (8) can also be written in terms of the thermal-neutral voltage as:

$$q"_R = i(V_N - V_{tn}) \tag{12}$$

Since the enthalpy of reaction, ΔH_R, is strictly a function of temperature (ideal gas approximation), the thermal-neutral voltage is also strictly a function of temperature, independent of cell operating pressure, *ASR* and gas compositions. The particular values of net cell heat flux at other operating voltages do however depend on cell *ASR* and gas compositions. The thermal-neutral voltage increases only slightly in magnitude over the typical operating temperature range for solid-oxide cells, from 1.287 V at 800°C to 1.292 V at 1000°C. At typical solid-oxide electrolysis stack temperatures and *ASR* values, operation at the thermal-neutral voltage yields current densities in the 0.2 – 0.6 A/cm² range, which is very close to the current density range that has yielded successful long-term operation in solid-oxide fuel cell stacks.

Operation at or near the thermal-neutral voltage simplifies thermal management of the stack since no significant excess gas flow is required and component thermal stresses are minimized. In fact, in the electrolysis mode, since oxygen is being produced, there is also no theoretical need for air flow to support the reaction at all. In a large-scale electrolysis plant, the pure oxygen produced by the process could be saved as a valuable commodity.

There are some additional thermodynamic implications related to the thermal neutral voltage. In particular, electrolyzer operation at or above the thermal neutral voltage negates the argument that is often stated as a motivation for high-temperature electrolysis that a fraction of the total energy requirement can be supplied in the form of heat. In fact, for isothermal operation at voltages greater than thermal neutral, heat rejection is required.

Electrolysis efficiency, η_e, can be defined for HTE, analogous to the definition of fuel cell efficiency[iv]. The electrolysis efficiency quantifies the heating value of the hydrogen produced by electrolysis per unit of electrical energy consumed in the stack. Based on this definition,

$$\eta_e = \frac{\dot{N}_{H_2} \Delta H_R}{VI} \tag{13}$$

and since the stack electrical current is directly related to the molar production rate of hydrogen via Faraday's law, the electrolysis efficiency can be expressed strictly in terms of cell operating potentials as:

$$\eta_e = \frac{\Delta H_R / 2F}{V_{op}} = \frac{V_{tn}}{V_{op}}. \tag{14}$$

The efficiency for the fuel-cell mode of operation is the inverse of Eqn. (14). A fuel utilization factor is often included in the fuel-cell efficiency definition, but it is not needed in the electrolysis definition since no fuel (only steam) is wasted at low utilization.

It should be noted that the value of the efficiency defined in this manner for electrolysis is greater than 1.0 for operating voltages lower than thermal neutral. As an example, for the reversible standard-state reference case, from Eqn. (4), on a rate basis:

$$\dot{W}_{rev} = \dot{N}_{H_2} \Delta G_R^o = IV^o \tag{15}$$

Invoking Faraday's law, the operating cell potential for this case approaches the reference open-cell value, $V^o = \Delta G_R / 2F$, yielding:

$$\eta_{e,o} = \frac{\Delta H_R^o}{\Delta G_R^o} \tag{16}$$

which for steam electrolysis at 850°C is equal to 1.34. For cases with variable gas composition or partial pressure, the open-cell potential is given by the Nernst Equation (6) and the corresponding efficiency limit varies accordingly. It is not desirable to operate an electrolysis stack near the maximum efficiency limit, however, because the only way to approach this limit is to operate with very low current density. There is a trade-off between efficiency and hydrogen production rate in selecting an electrolysis stack operating voltage. This trade-off is illustrated in Fig. 9. The upper curve in the figure shows the decrease in electrolysis efficiency that occurs as the per-cell operating voltage is increased above the open-cell voltage, V_N, according to Eqn. (14). Operation at the thermal-neutral voltage yields an electrolysis efficiency of 1.0 (i.e. 100%). Area-specific resistance (*ASR*) represents the net effect of all the loss mechanisms in the electrolysis stack including, ohmic losses, activation and concentration overpotentials, etc. The bottom curves show the effect of operating voltage and *ASR* on the current density. Noting that:

$$V_{op} = \overline{V}_N + i \times ASR, \tag{17}$$

if a target current density (and corresponding hydrogen production rate) is selected, lower *ASR* values allow for stack operation at lower voltages and correspondingly higher efficiencies. A good tradeoff between efficiency and hydrogen production rate will occur at operating voltages below $\Delta H_R/2F$, around 1.1 V. The challenge is to develop SOEC stacks with low *ASR* such that a reasonable current density will be achievable at lower operating voltages.

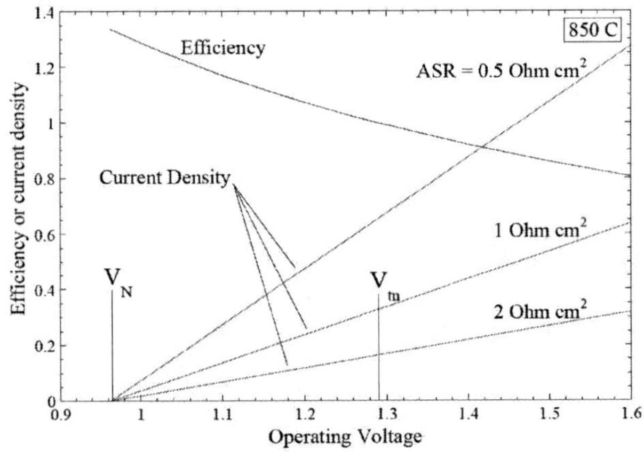

Figure 9. Electrolysis cell efficiency and current density as a function of operating voltage and cell resistance.

Co-electrolysis is essentially a variation of high temperature steam electrolysis. An electrical potential is applied across a gas tight and electrically insulating ceramic membrane, having a high conductivity of oxygen ions. If the potential is greater than the free energy of formation, adjusted for local temperatures and reactant and product partial pressures, an H_2O or CO_2 molecule will decompose as one oxygen atom is transported across the membrane in the form an oxygen ion ($O^=$) leaving behind hydrogen or carbon monoxide. However, quantitative analysis of co-electrolysis is significantly more complex than simple steam electrolysis.

This is primarily due to the multiple, interacting reactions that occur: steam electrolysis, CO_2 electrolysis, and the reverse shift reaction (RSR):

$$CO_2 + H_2 \leftrightarrow CO + H_2O, \tag{18}$$

Reaction kinetics govern the relative contributions of these three reactions. It is also important to note that the electrolysis reactions are not equilibrium reactions since the electrolyte separates the products from the reactants. However, the RSR is a kinetically fast, near equilibrium reaction at high temperature in the presence of a Ni catalyst. Further, the electrolysis cell cathode (anode in fuel cell operation) is typically a nickel ceramic composite and an effective shift or reforming catalyst. Note that all four species participating in the RSR are present on the cathode (hydrogen electrode in SOEC mode) as shown in Figure 10.

Feed: H_2O, CO_2, (minor H_2, CO)

Reverse Shift Reaction: $CO_2 + \Uparrow H_2 \Longleftrightarrow CO + \Downarrow H_2O$

As steam is consumed and H2 produced the RSR proceeds to the right

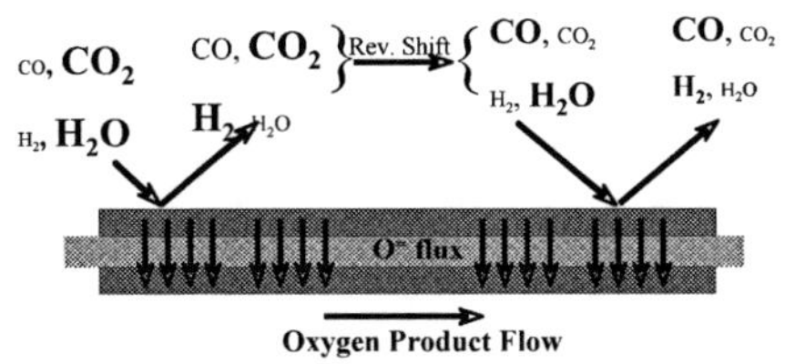

Figure 10. Schematic of high temperature co-electrolysis

Also, if the cell potential is high enough, CO can be further electrolyzed to elemental C:

$$CO \rightarrow C + 1/2 O_2, \tag{19}$$

producing solid deposits on cell surfaces that degrade cell performance. The reduction potential of CO is usually only slightly higher than that of CO_2, and under some combinations of temperature and p_{CO}/p_{CO2} is actually lower (Figure 11). Additionally, the electrochemical kinetics of dry CO_2 electrolysis are worse than steam electrolysis, resulting in cell resistances from 1.5 to 10 times higher (depending on the electrode material set) than observed in steam electrolysis. While it is possible to produce syngas by separately electrolyzing steam and CO_2, there are significant advantages to electrolyzing steam and CO_2 simultaneously, primarily lower cell resistance and reduced potential for carbon formation by CO electrolysis.

In co-electrolysis, the reverse gas shift reaction is thought to be responsible for most of the CO production, as the overall cell resistance matches that observed for pure steam electrolysis. When the shift reaction is in equilibrium, the calculated local cell Nernst potential given by Eqn (20) for CO_2 electrolysis will be identical to that given by Eqn (6) for steam electrolysis. Therefore, there is no thermodynamic potential advantage for steam electrolysis over CO_2 electrolysis, only the electro-kinetics advantage.

$$V_N = V^o - \frac{R_u T}{jF} \ln \left[\frac{y_{CO2}}{y_{CO} y_{O2}^{1/2}} \left(\frac{P}{P_{std}} \right)^{-1/2} \right] \tag{20}$$

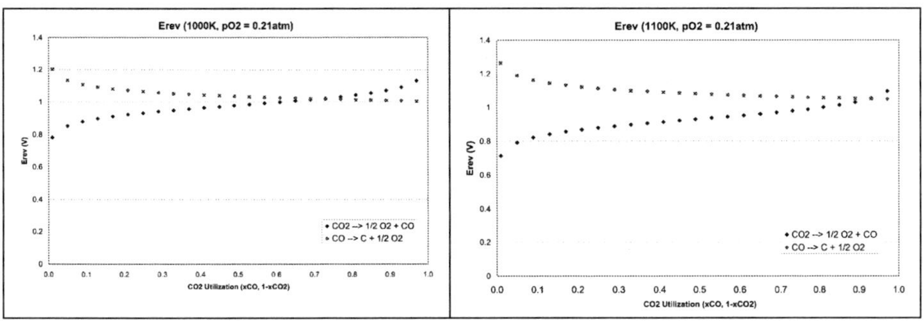

Figure 11. Equilibrium reduction potentials of CO and CO_2 in dry CO-CO_2 mixtures.

High Temperature Electrolysis Testing

The cell polarization curves shown in Figure 12 are of one cell, first tested in fuel cell mode with humidified hydrogen to determine the baseline cell resistance, then tested in both electrolysis and fuel cell modes with a dry CO/CO_2 mixture. The cell resistance was 50% higher while operating on dry CO/CO_2 as compared with humidified hydrogen. Another cell, operated in both co-electrolysis and fuel cell modes with a humidified CO-CO2 feed stream, showed resistance identical to its performance in pure steam electrolysis mode (Figure 13). The effect of steam on CO_2 electrolysis electro-kinetics was clearly demonstrated in Figure 14, where a cell under a constant voltage condition of 1.2V was subjected to a change in feed composition from a mixture containing hydrogen and steam along with the CO/CO_2 to a feed containing only dry CO/CO_2. It can be seen that as the hydrogen and steam are swept from the cell chamber over a few minutes (at ~0.85 hours on the graph) the cell current drops and the conversion of CO_2 to CO drops proportionally. It appears than only a small amount of H_2/H_2O is required to sustain the low cell resistance.

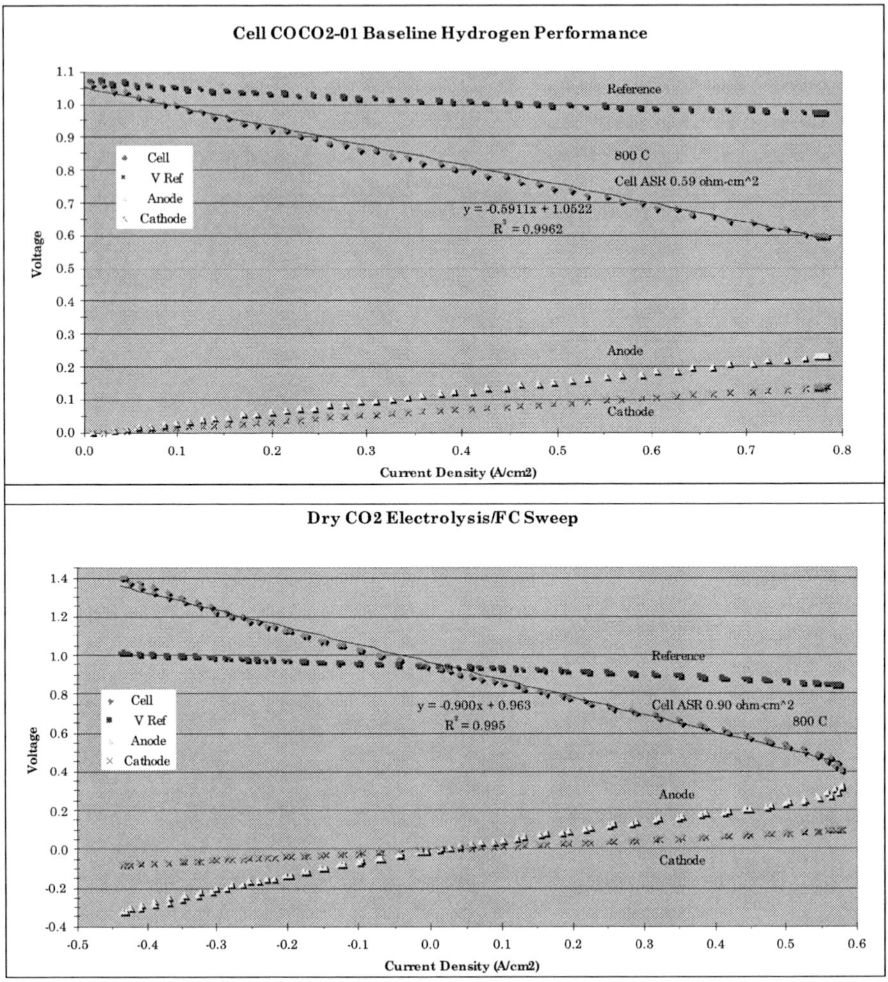

Figure 12. Cell resistance change between humidified hydrogen and dry CO-CO_2 mixtures.

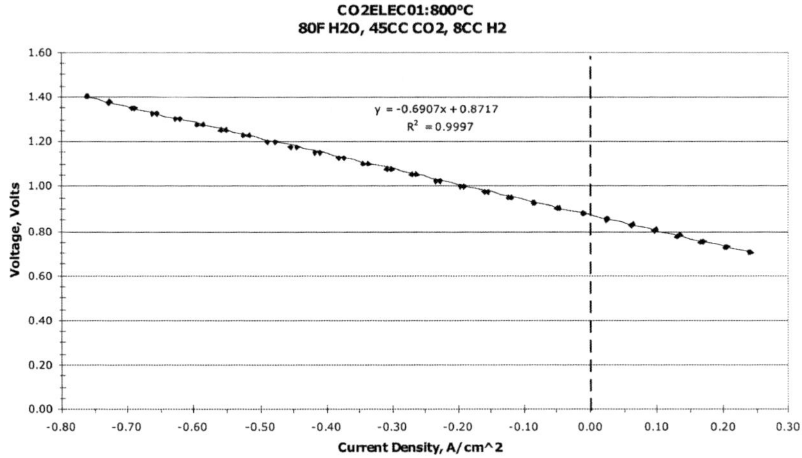

Figure 13. Co-electrolysis cell showing unchanged resistance with humidified CO-CO_2 mixtures.

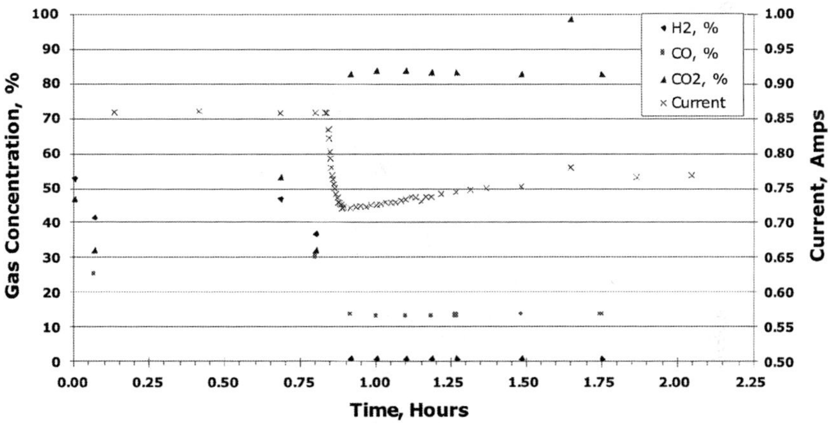

Figure 14. Co-electrolysis composition and current change from humidified to dry CO-CO_2 feed at 0.85 hours.

ILS Test at the INL

A 3-module, 12 stack, 720 cell Integrated Laboratory Scale (ILS) test was conducted at the Idaho National Laboratory as part of the DOE Nuclear Hydrogen Initiative. The ILS was operated for 1000 hours in the fall of 2008. Peak hydrogen production rate at the thermal neutral voltage was 5.7Nm^3/hr with an electrical input power of 18kW (Figure 15).

Each of the four stacks comprising a module were connected electrically in parallel using electrical tie bars welded to interconnect tabs after every fifth cell. Each stack had its own set of current collecting buss-bars welded to the top and bottom interconnects tabs. These buss-bars exited through the bottom of the test furnace and were connected to the external power sources. Voltage taps were inserted into the air exhaust channels of one of the stacks every 10th cell. Three internal thermal couples were also inserted into each of the four stacks at locations near the top, middle, and bottom of the stack. Each stack consisted of 60 electrolyte-supported cells that were 10 cm x 10 cm outside dimensions and had an active area of 61 cm2 per cell. The electrolyte was scandia-stabilized zirconia, approximately 220 microns thick. The hydrogen electrode was Ceramatec's standard nickel cermet composite. The hydrogen electrodes were also printed with unfired nickel ink as a current distribution layer. The air electrode was Ceramatec's standard manganite perovskite formulation, with an unfired lanthanum strontium chromite (LSC) current distribution layer.

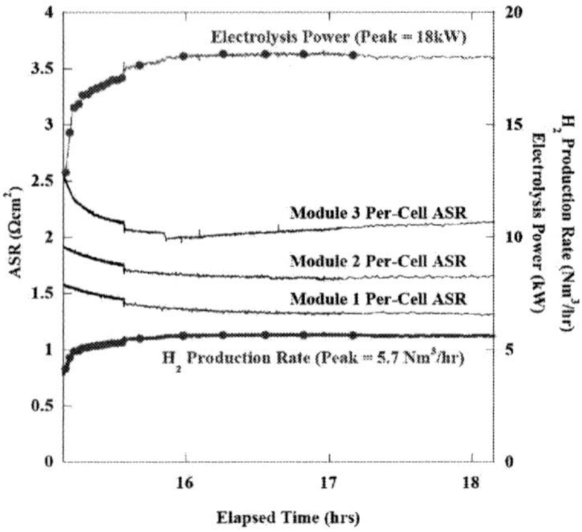

Figure 15. Hydrogen production rate during the Integrated Laboratory Scale electrolysis test at the INL.

The interconnects employed treated metal components in a cross-flow configuration. The separator plate was 400-series stainless with a rare earth scale stabilization treatment and an LSC current distribution layer applied on the oxygen evolution electrode side. The oxygen side flow channel was nickel-chrome alloy corrugation with the scale stabilization treatment and was coated with an LSC layer. Ferritic stainless edge rails were also treated with a ceramic coating to inhibit steam corrosion and were glued in place for assembly. The hydrogen flow channel was an untreated nickel foil corrugation. The stack was assembled using the above components in an alternating pattern.

Conclusions

High temperature co-electrolysis provides a flexible and highly efficient technology for converting CO_2 to transportation fuel feedstock. The electrolyzer can produce hydrogen for conventional CO_2 conversion, or if more convenient from an overall process standpoint the CO_2 can be converted directly in the electrolysis step. Energy from a variety of renewable generation technologies can be used to power the electrolyzer. The resulting synfuel is a means of storing renewable energy in a way that adds considerable value over wholesale electric grid pricing. Economic projections, assuming $50/MW-hr electric power costs, and SOFC system capital costs of $400/kW (DOE SECA targets), suggest that this process is not competitive with current $70/bbl oil, but may be if oil prices once again climb above $130/bbl. Implementation of some sort of carbon emissions cost would further close the gap by $10-$20/bbl.

Acknowledgements

This work was supported by the Idaho National Laboratory, Laboratory Directed Research and Development program, the U.S. Department of Energy, Office of Nuclear Energy, Nuclear Hydrogen Initiative Program, and the Office of Naval Research.

References

i Schimel, D., and Moore, B., data in *Trace Gases and the Biosphere*, UCAR, Boulder, CO., 1992

ii Hartvigsen, J. et al, "Carbon Dioxide Recycling by High Temperature Co-electrolysis and Hydrocarbon Synthesis", p 171-182, 2008, Carbon Dioxide Reduction Metallurgy - TMS 2008 Annual Meeting, New Orleans, LA.

iii Harvego, E. A., McKellar, M. G., Sohal, M. S., O'Brien, J. E., and Herring, J. S., "Economic Analysis of a Nuclear Reactor Powered High Temperature Electrolysis Hydrogen Production Plant," ASME 2nd International Conference on Energy Sustainability, Jacksonville, FL, August 10-14, 2008.

iv Hartvigsen, J., Elangovan, S., and Khandkar, A., "Selection of SOFC Stack Operating Point for Optimal Balance of Efficiency and Power", 1998, Third European Solid Oxide Fuel Cell Forum, Nantes France.

Energy Technology 2010
Edited by: Neale R. Neelameggham, Ramana G. Reddy, Cynthia K. Belt, Ann M. Hagni and Subodh Das
TMS (The Minerals, Metals & Materials Society), 2010

SUNSHINE TO PETROL: A METAL OXIDE-BASED THERMOCHEMICAL ROUTE TO SOLAR FUELS

James E. Miller[1], Richard B. Diver[1], Nathan P. Siegel[1], Eric N. Coker[1], Andrea Ambrosini[1], Daniel E. Dedrick[2], Mark D. Allendorf[2], Anthony H. McDaniel[2], Gary L. Kellogg[1], Roy E. Hogan[1], Ken S. Chen[1], Ellen B. Stechel[1]

Sandia National Laboratories
[1]PO Box 5800
Albuquerque, NM 87185, USA
[2] PO Box 969
Livermore, CA 94551-969, USA

Keywords: Metal oxide, Thermochemical cycles, Fuels

Abstract

Converting carbon dioxide and water to hydrocarbons is an attractive option for storing solar energy and, coupled with appropriate CO_2 capture technology, for recycling carbon and impacting atmospheric CO_2 concentrations. For any process, high solar-to-fuel efficiency is necessary for large scale viability and favorable economics. Thermochemical approaches for solar-to-fuel conversion are potentially highly efficient as they avoid the inherent limitations of photosynthesis and also sidestep the solar-to-electric conversion necessary to drive electrolytic reactions. Solar-driven two-step metal-oxide-based thermochemical cycles for producing the components of syngas, CO and H_2, from CO_2 and H_2O are the basis of the "Sunshine to Petrol" project. Multi-cycle production of both H_2 and CO has been demonstrated over several iron- and cerium-based compositions fabricated into monolithic pieces both in the laboratory and at the National Solar Thermal Test Facility. These compositions are being developed for deployment in a unique and continuous solar-driven reactor prototype, the counter-rotating-ring receiver reactor recuperator or CR5.

Introduction

The interconnected challenges posed by increasing global competition for petroleum and the increasing concentration of atmospheric carbon dioxide are now widely recognized. Meeting these challenges requires recognition that although hydrocarbon fuels are ideal energy carriers, they can no longer be thought of as primary energy sources. It is necessary to take the realistic view that our conventional fossil fuels are in fact "stored sunlight" in the form of "sequestered carbon." This view leads to a search for options to use and store sunlight more directly.

Sunshine to Petrol

In combusting hydrocarbons we unlock the energy for utilization, but we also release the carbon to the atmosphere. However, nature teaches us that with an appropriate persistent energy source, e.g. solar irradiation, it is possible to effectively reverse the combustion process and "re-energize" CO_2 and H_2O back into hydrocarbon form.

$$\text{Solar Energy} + xCO_2 + (x+1)\,H_2O \rightarrow C_xH_{2x+2}(\text{liquid fuel}) + (1.5x+.5)\,O_2$$

The completion of this cycle (Figure 1) by biological means broadly defines the biofuels approach. The basic proposition of "Sunshine to Petrol" (S2P) is that a more direct route to hydrocarbon fuels, e.g. through the application of chemical technology, offers many potential advantages. Through this lens, the above equation can be seen as an extension of the "hydrogen economy" through the inclusion of carbon (CO_2). This extension captures many of the benefits of the proposed hydrogen economy while also preserving the benefits and much of the infrastructure of the existing hydrocarbon economy. At the heart of the S2P approach is metal oxide-based cyclic thermochemical conversion of CO_2 and H_2O to CO and H_2, the basic building blocks for synthetic fuels [1].

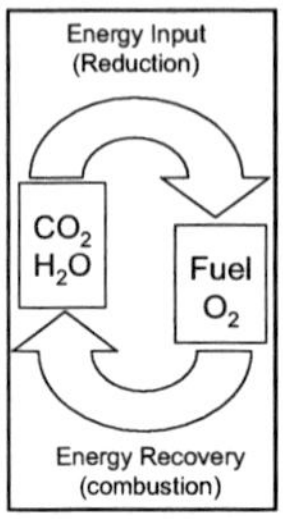

Figure 1. The vision of "Sunshine to Petrol" is to apply chemical and materials technology to directly "reverse combustion" by "energizing" CO_2 and H_2O. This vision is analogous to the biofuels approach, and also an extension of the "hydrogen economy".

Why Target Fuels for CO_2 Utilization?

CO_2 capture from industrial sources and even ambient air is an area of active research and development, and interest is growing in productive utilization of CO_2 as an alternative to sequestration. In addition to serving as a feedstock for manufacturing synthetic fuels, CO_2 could be utilized on an industrial scale for a number of other purposes. For example, CO_2 can serve as a feedstock for manufacturing chemicals, or as a solvent or reaction media [2,3,4]. Although this chemistry can be very elegant, the relatively small volumes that would be used render these approaches almost irrelevant from an atmospheric CO_2 point of view. The only potential product family that could consume CO_2 feedstock at a level commensurate with current emissions is hydrocarbon fuels, the original source of much of the CO_2.

Table I. Potential Impact of Using CO_2 as a Feedstock for Chemical Production.

	2004 U.S. production [5] (1,000 metric tons)	CO_2 equivalents (1,000 metric tons)
Ethylene	25,682	80,641
Propylene	15,345	48,183
Ethylene dichloride	12,163	10,811
Sum of Top 3 U.S. Chemicals		139,635
U.S. Petroleum [6]		2,598,000
Petroleum, Coal, NG [6]		5,912,000

Table I illustrates this point. Ethylene, propylene, and ethylene dichloride are three largest commodity chemicals produced in the United States. The third column of the table shows the amount of CO_2 that could be consumed annually as a feedstock to produce these chemicals assuming 100% carbon efficiency. For example, the carbon in one mole of ethylene (C_2H_4) or ethylene dichloride ($C_2H_4Cl_2$) could theoretically be provided by 2 moles of CO_2; thus one pound of ethylene could theoretically consume 3.14 pounds of carbon dioxide. The 2004 emissions of CO_2 in the United States tied to fossil energy sources are provided in the bottom rows. A comparison shows that a CO_2 utilization scheme for chemical production, although potentially valuable, would account for only a small fraction of the CO_2 being produced. The CO_2 equivalents of the three leading chemicals amount to only 5.4% of the CO_2 resulting from combustion of petroleum products and only 2.4% of that tied to petroleum, coal and natural gas.

Utilization of CO_2 as polymer precursors does have the advantage of essentially being a sequestration mechanism. In contrast, converting industrial sources of CO_2 (flue gas) into fuels that will eventually be combusted represents a "second-use" of the carbon that does not directly sequester carbon but rather slows the net rate of CO_2 release by an amount that is proportional to the stored solar energy. Eventually reducing the net rate of CO_2 release to zero will require a viable air or on-board capture mechanism. A full life cycle analysis remains to be completed.

Efficiency Matters

The United States currently consumes about 20 million barrels of petroleum each day (mbpd), while global demand exceeds 80 mbpd. The efficiency at which this oil was produced from sunlight is believed to be quite low, and has been estimated as only 0.0002% [7]. At this efficiency, the sunlight falling on a square meter of Albuquerque, NM (2600 $kWh/m^2/yr$) for 300,000 years would produce less than a single barrel of oil (6120 MJ/barrel). Of course, very long time scales have resulted in an accumulation of vast quantities of petroleum that allow current demands to be met. Petroleum alternatives, i.e. solar fuels, face the challenge of meeting or offsetting a significant portion of petroleum demand on a much shorter timescale. Thus, much higher efficiencies are required.

For biofuels, the overall sunlight to fuel efficiency is dependent on location and the process specifics and is thus difficult to define precisely or to generalize. However, one can put an upper limit on the biomass approach by considering the efficiency of the photosynthetic step alone. Actual efficiencies will be much less. The maximum possible efficiency is estimated to be 4.6% and 6% for C3 and C4 photosynthesis respectively under current atmospheric conditions [8]. Generally though, photosynthesis is measured to be only 2.5% efficient at best [7], and is less than 0.5% for rapidly growing large area crops [9]. To put these efficiencies into perspective, one can calculate a land area require to produce 20 mbpd oil equivalent for a given solar resource and solar-to-fuel efficiency. Doing so with the generous assumption of uniform irradiance across the area at an average rate of that realized in Albuquerque, NM leads to the conclusion that an area roughly equal to the entire western half of the continental U.S. is required for an efficiency of 0.1%. Clearly, even though the efficiency is comparatively high relative to fossil fuels, in the absence of very significant advances, this approach can be only a partial solution due to real-world resource limitations.

Given the limits on overall sunlight-to-hydrocarbon efficiency imposed by photosynthesis, it is reasonable to consider conventional synthetic chemical approaches. Hydrogen can currently be produced renewably with commercially available technologies, e.g. via photovoltaic-driven electrolysis. A reasonable starting estimate for the solar-to-hydrogen efficiency is about 9% (0.12(PV) x 0.75(electrolysis) = 0.09). An actual efficiency of 6.2% has been reported for long term operation of a PV driven electrolytic system [10]. Hydrogen may be reacted with CO_2 to directly produce methanol, or indirectly to produce CO and then methanol, for example. Plant models have been used to predict that current technology would allow methanol to be manufactured from CO_2 and electrolytic H_2 with an efficiency of 49% [11]. Others estimates in the range of 40-50% electricity to fuel (included carbon capture) have also been published [12,13]. Hence, sunlight-to-fuel efficiencies of at least 3% are plausible for PV-driven fuel production processes. Following the earlier example, at this efficiency, PV collectors with an area totaling the size of the state of Georgia would be required to produce 20 mbpd equivalent.

In examining the electrolytic hydrogen approach to solar fuels, it is clear that the biggest single efficiency loss arises from the conversion of photons to electric power. Herein lies the motivation for developing thermochemical processes for H_2 and CO production from H_2O and CO_2. Concentrating solar flux and using it to drive thermochemical processes effectively uses more of the solar spectrum and avoids the costly solar to electric conversion.

Basic Cost Considerations

The economics and optimization of the overall S2P process in particular, and solar fuels in general, is an area of active consideration by the project team, see for example [11]. Detailed results and discussion will not be presented here. However, one important conclusion bears mentioning. Under almost any scenario, the largest contributor to the final cost of chemically producing a solar fuel will be in the initial, most-energy-intensive solar step. That is, in our case, the conversion of the stable combustion products CO_2 and H_2O into synthesis gas will be the predominant contributor to cost.

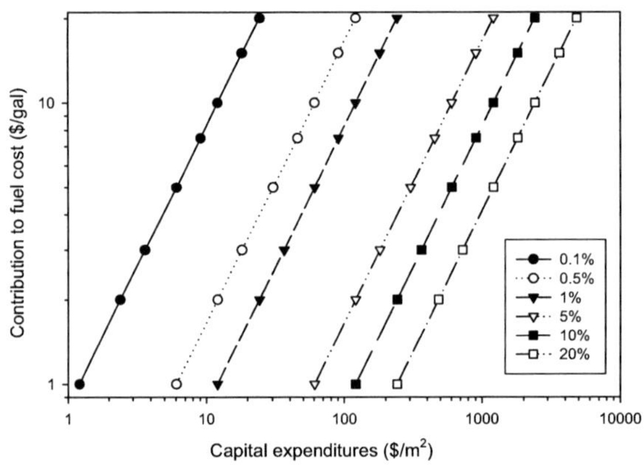

Figure 2. Contribution to fuel production cost of capital expenditures for solar capture and conversion as a function of overall (solar-to-fuel) system efficiency. Assumptions, gge = 36 kWh, solar resource = 2600 kWh/m²/yr, 30 year amortization, 5% interest.

As the solar cost is key, the efficiency and the cost of each increment of added efficiency is likewise key. Figure 2 shows an example of the contribution to the fuel cost attributable to capital expenditures for solar collection and conversion for a given overall system efficiency. For the purposes of this discussion, capital expenditures are assumed to be paid annually over a 30 year period with 5% interest. From the figure, if the impact on cost is to be kept below \$5/gal, then solar cost must be kept below \$6/m^2 if the system is 0.1% efficient, or \$600/m^2 if the system is 10%. Note that costs of supporting systems and balance of plant tend to become more significant as the efficiency decreases and size increases. A land cost of \$1/m^2 is equivalent to only \$4050/acre. Feedstock costs are relatively small. Estimates of the cost for CO_2 recovery from air vary, but experimental results support an ultimate cost of \$200/ton or less [14]. This price contributes less than \$2/gal to the production cost, assuming 100% carbon efficiency.

To provide some perspective, the 2007 installed cost for large scale PV was reported to average \$6.80/W [15]. As PV costs per Watt are fairly independent of efficiency [15,16] this is roughly \$680/m^2 for a 10% efficient module, which we can estimate from the discussion above as giving a 5% system efficiency. This contributes over \$11/gal to the price of fuel production, and is an optimistic assumption as it is based on peak numbers. In contrast large parabolic dish concentrators cost roughly \$300/m^2, and have a potential for a higher overall efficiency.

Thermochemical Cycles for Water and Carbon Dioxide Splitting

Carbon dioxide and water are extraordinarily stable molecules. Thermodynamics does not favor their decomposition, even at 3000 °C. In addition to very high temperatures, direct thermolysis is also made difficult in a practical sense by the inherent spatial coupling of the evolution of the different reaction products (CO and O_2, or H_2 and O_2). Thermochemical cycling is an approach for overcoming these challenges wherein an unfavorable, but desired, reaction is divided into two or more thermodynamically favorable reactions that net the first. Metal oxide cycles are particularly attractive as they are simple (they require only two steps), and feature gas solid/reactions for easy isolation of the working solid material from the desired products. However, the very high temperatures required limit their applicability to being driven by concentrating solar power (CSP) only, more specifically parabolic dish concentrators.

The basic concept is illustrated using a ceria-based two-step carbon dioxide splitting (CDS) cycle as an example in Figure 3. Other metal oxide cycles, e.g. ferrite cycles, are conceptually identical. Briefly, CeO_2 is heated to high temperatures causing thermal reduction to occur. The product oxygen is pumped or swept away leaving a reduced oxide. The figure shows full reduction to Ce_2O_3; partial reduction to intermediate phases or mixtures of phases is the more general case. At lower temperatures, the reduced ceria is now thermodynamically oxygen deficient. As such, it can be reoxidized with CO_2 to remake CeO_2 and yield CO.

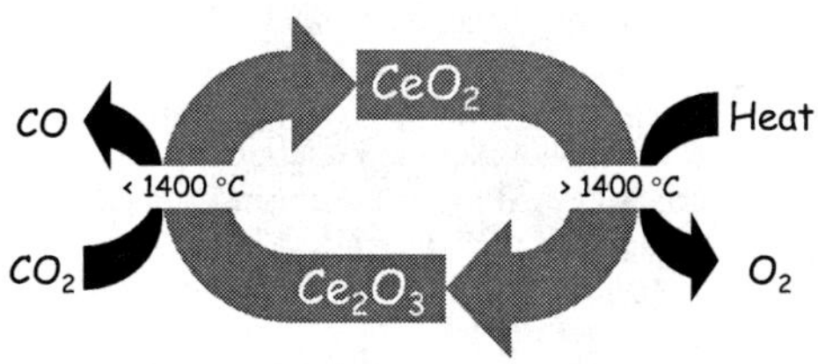

Figure 3. Conceptual diagram of ceria-based thermochemical cycle. For discussion of thermodynamics of metal oxide thermochemistry, see [17,18].

Water splitting (WS) is entirely analogous to carbon dioxide splitting. Water splitting cycles have been reviewed elsewhere [19]. One important difference however is that at temperatures above about 800 °C, CO_2 is less stable than H_2O, while at lower temperatures it is more stable. This results in oxidation of the reduced metal oxide (left side of Figure 3) being more favorable with CO_2 than H_2O at temperatures greater than 800 °C. Improvements in the efficiency of downstream processes (fuel synthesis) may also result [11,20].

The Counter-Rotating-Ring Receiver/Reactor/Recuperator

One of the principal advantages of thermochemical cycles is that they avoid the efficiency-sapping sunlight-to-electrical energy conversion required for electrolytic processes and therefore have the potential to improve the overall efficiency of renewable syngas production. However the thermal cycling required by thermodynamics may also introduce large efficiency losses into the system unless the heat is recuperated between cycles. For example, it is readily shown that in a hypothetical case of a Fe_3O_4/FeO system operating at temperatures of 2027 °C (reduction) and 327 °C (reoxidation), the maximum thermal to chemical efficiency that could be obtained is only 36%. However if heat is recovered from the cooling step and applied to reheat the working oxide, the maximum possible efficiency is increased to 76% [21]. With this in mind, we are currently developing a novel solar-driven device for splitting CO_2 and H_2O based on the philosophy of "reactorizing a counter-current recuperator." We call this heat engine the Counter-Rotating-Ring Receiver/Reactor/Recuperator, or CR5 [21].

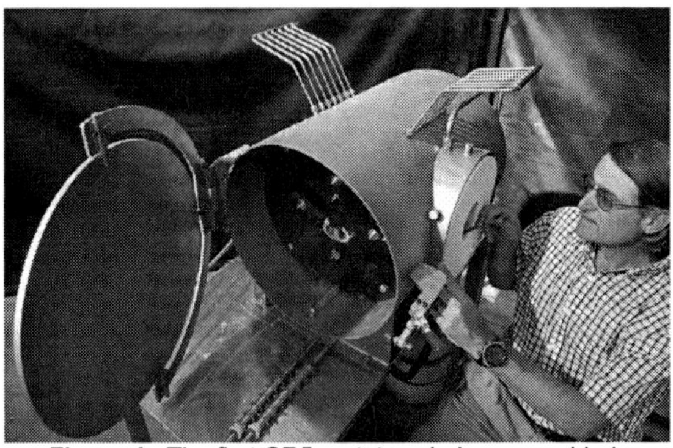

Figure 4. The first CR5 prototype being assembled.

The key feature of the CR5 is stack of counter-rotating rings (rotation speed ~ ½ rpm) with fins constructed of a reactive metal oxide positioned around the circumference. The rings rotate continuously through a section illuminated with concentrated solar flux wherein they are reduced. A cool zone lies opposite to the illumination area wherein oxidation occurs. Linking these two zones are two sections wherein thermal radiation transfers sensible heat between the counter-rotating rings. In addition to providing a mechanism for internal recuperation of heat, the rotation of the reactive rings within the CR5 provides a mechanism for continuous operation. That is, no sunlight is wasted as material continuously passes through the illuminated region.

Materials Challenges

The viability of the S2P vision for hydrocarbon fuels hinges on the economics and scalability, particularly of the metal oxide thermochemistry. In turn, the viability of the CR5 approach hinges on the thermal efficiency and durability of the materials. Chemical and structural requirements must simultaneously be met. For example, rapid kinetics for both reactions (reduction and oxidation) enable greater mass and volumetric utilization of the materials, and hence improve efficiency. This drives one towards metal oxide structures that maximize surface to volume ratios. Fine feature sizes will maximize the gas/metal oxide interface. The solar/metal oxide interface is equally important. The short cycle times dictated by heat transfer considerations severely limits thermal conduction. Hence, material that is not directly illuminated is unlikely to reach optimum reduction temperatures. Thus the ideal structure will be an open, high surface are "volumetric absorber." However, the structure must also maintain its physical integrity through rotation, and thermal and chemical cycling. Hence the structure should be robust and stresses created by thermal and chemical changes should be minimized. The S2P project has several thrusts aimed at moving materials forward to meet these and related challenges. A brief introduction to our activities in several materials topic areas follows.

Materials Validation

Early work led us to focus on substituted ferrites, particularly cobalt-substituted ferrites (CFO) in a yttria-stabilized zirconia (YSZ) matrix as a baseline reactive oxide [22]. Thermodynamic and other considerations that favor using substituted ferrites over the simpler iron-only system are presented elsewhere [18]. Our efforts also include ceria-based materials [17,23]. The CR5 requires that the reactive oxide materials, typically developed and studied as powders in the laboratory, be converted into solid monoliths that can be mounted on the circumference of a ring. Our approach has been to directly fabricate the powders into monolithic structures, rather than affix the powders to a monolithic support [17]. In doing so, we avoid problems such as thermal mismatch between substrate and powder coating and reactions between reactive oxide and the support.

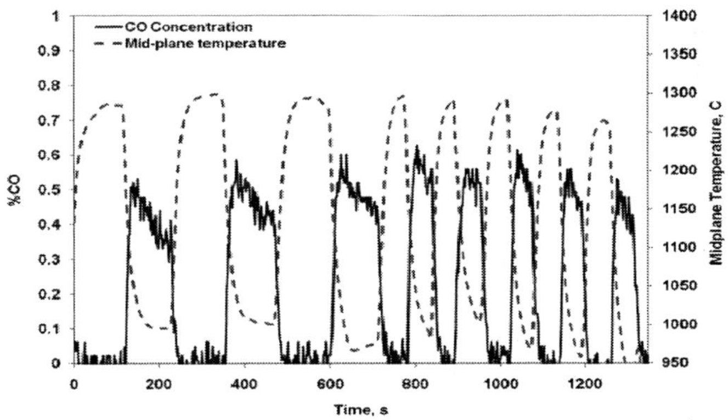

Figure 5. Solar-driven CO_2 splitting thermochemical cycles using $Co_{0.67}Fe_{2.33}O_4$/YSZ composites (1:4 by weight). Half cycle times were two minutes (left) and one minute (right). Dotted line shows temperature, solid line shows CO concentration downstream of reactor. Oxygen evolution data not shown.

Initially, there were questions as to whether this approach would be feasible as bulk ferrites are known to be unreactive over multiple cycles. Several different monolithic forms of materials, including pressed pellets and lattice structures fabricated via the robocasting technique [24], were validated both in the laboratory and in solar reactors at the National Solar Thermal Test Facility (NSTTF). For solar testing, samples were held in a stationary position and temperatures were cycled by controlling solar flux, while gasses flowing through the monolith were cycled between inert (thermal reduction) and CO_2 (or humidified inert) [25]. The success of the direct fabrication approach for is demonstrated for CFO/YSZ composites in Figure 5.

Materials Characterization

Materials characterization efforts are structured to meet two critical ends, designing and optimizing metal oxide working materials, and developing the quantitative knowledge and models of reactions and transport processes required to aid in the design of future generations of the CR5. Towards the first end our efforts are currently focused on gaining detailed understanding of the ferrite systems. To simplify the complex system, much of the characterization has focused on mixtures of unsubstituted iron oxides with YSZ. A number of observations provide motivation including 1) a predominant zirconia or hafnia phase is required for ferrite monoliths to function 2) the composite systems evolve over time; aged materials show higher yields and in some cases improved kinetics 3) reaction rates and orders are different for CO_2 and H_2O, and 4) combinations of CO_2 and H_2O exhibit synergistic behavior (Figure 6).

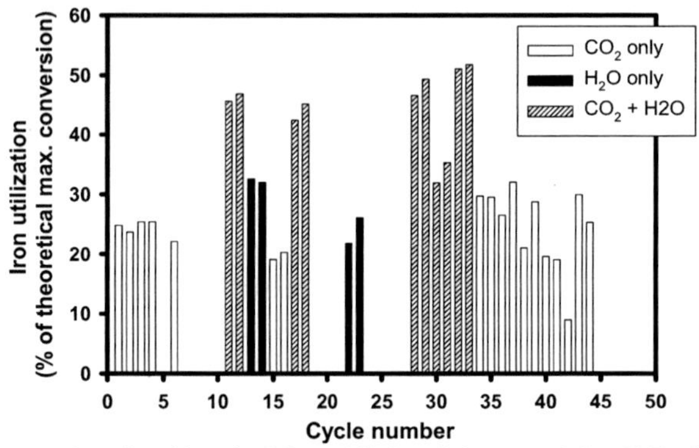

Figure 6. Demonstration of multi-cycle CO_2 and H_2O splitting over 4.8 wt% Fe_3O_4 in YSZ. For the cycles shown, thermal reduction was carried out at 1450 °C and oxidation at 1100 °C. Note enhanced yields (Fe utilization) when CO_2 is humidified for the reoxidation step. Fe utilization is calculated relative to all Fe participating in the theoretical cycle $Fe_3O_4 \rightarrow 3FeO$, $3FeO + CO_2 (H_2O) \rightarrow Fe_3O_4 + CO (H_2)$.

In-situ X-ray diffraction of Fe/YSZ samples while temperatures and gas environments are cycled have provided insights into the evolution of the material over time [26]. These studies have been augmented by thermogravimetric analysis, electron microscopy, and bulk diffusion/reactivity studies to name a few. It is clear that Fe is soluble in the YSZ phase, but more importantly the behavior and degree of solubility varies depending on the loading, temperature, and the oxidation state of the iron. In a working system, the Fe appears to be quite mobile and readily moves into and out of the YSZ "support." The utilization of soluble Fe appears to be quite high relative to the bulk ferrites, particularly relative to Fe_3O_4. Cobalt substitution increases the utilization of the bulk materials.

An example of kinetic data collected for reactor modeling and design purposes is shown in Figure 7 for a sample of nominal composition $Ce_{0.4}Zr_{0.6}O_2$ with 5% NiO (w/w). Ceria/zirconia mixed oxides are of interest as this modification is expected to improve stability and transport/reaction within the bulk. Thermochemical CO_2 and H_2O splitting has been demonstrated for > 60 cycles with some Ce/Zr formulations. The figure demonstrates that CO_2 can be split over the material with an activation energy of about 32 kcal/mol. At high temperatures, the rate saturates, either due to thermodynamic or transport limitations. As for ferrites, humidifying the CO_2 increases the reaction rate; the activation energy is unchanged. WS is thermodynamically unfavorable at temperatures above 1100 °C. The WS reaction has a first order concentration dependence, while CDS exhibits a 0.7 order dependence at 1250 °C (data not shown). Notably, the extent of reduction of the ceria showed relatively little dependence on the Ce/Zr ratio. Furthermore, the oxidation reactions are significantly slower than the reduction reaction. Pt addition significantly increases the reaction rate, but Pt is not stable at the high thermal reduction temperatures. Work to identify high-temperature-stable promoters is ongoing.

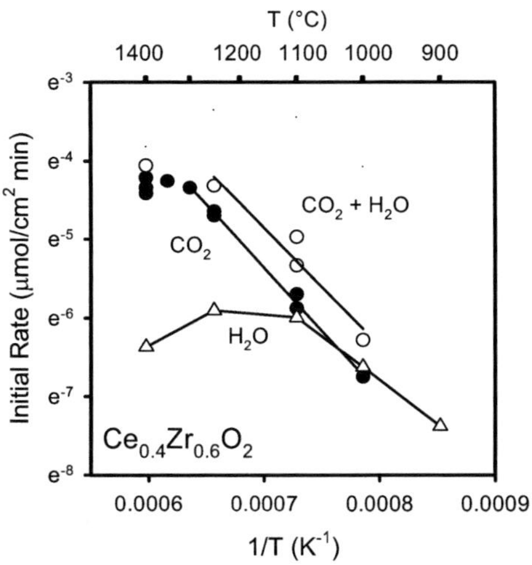

Figure 7. Reaction rates normalized by surface area for reoxidation of textured ceria/zirconia monolith thermally reduced for one hour at 1450 °C. Basic geometry of test sample shown below.

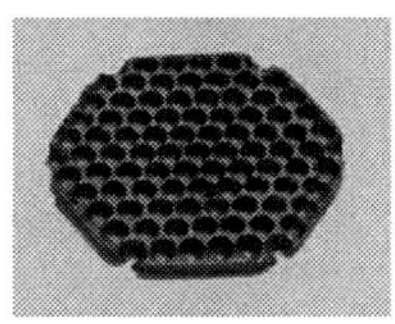

Models at the local materials scale and larger CR5 scale are being developed; initial results are discussed elsewhere [27,28].

Structural design, Characterization, and Validation

Several scalable fabrication processes and geometries that could meet the solar interface requirements are being developed in parallel. To date, we have identified four promising structures for evaluation. Thin plate geometries are fairly straightforward structures to produce and can be made with a relatively high surface to volume ratio that can be further increased by texturing the surface itself (Figures 7 and 8A). Foams (8B) have been used in the past for solar reactors. Their main advantage is that they can have a very high surface to volume ratio to support chemical reactions. Their main disadvantage is a susceptibility to mechanical failure at the "necking" points within the material.

The vertical pin structure (8C) consists of numerous pins attached at the base and arranged such that the concentrated sunlight is incident along the axis of the pins. This type of geometry has the advantage of being easily modified with respect to pin diameter, cross section, height, and spacing making it a particularly good platform for design optimization. The textured plates and pins in Figure 8 were fabricated with a technique developed at Sandia that involves mixing ceramic powders with an epoxy binder to form readily moldable material. The binder that renders the green material moldable burns away during the high temperature sintering of the final part [25]. A fourth geometry under investigation is ceramic felts (not shown). The principal advantage of these fiber mats is their high geometric surface areas; potential modifications to decrease their fragility are under investigation.

Each of the four candidate structures is being subjected to short-term on-sun and laboratory testing. Following this initial qualification, promising candidate structures and materials will be subjected to up to 1000 thermal and chemical cycles in an automated system with periodic evaluation of chemical and physical changes in the materials. Monolith design and integration of the monoliths into rings is being further aided by computational means, e.g. by analyzing stresses resulting from thermal gradients.

Figure 8. Composite Fe/YSZ monolithic test structures A) textured plate mounted for on-sun testing, B) ceramic foam, and C) vertical pin.

Another concern for durability is the potential for volatilization and subsequent loss of the reactive material, particularly for the ferrites. Both thermodynamic modeling and experiment are being used to investigate this possibility. Thermodynamic models indicate that metals such as Co and Ni that are commonly substituted into ferrites to improve their behavior are more volatile than the Fe. Steam increases the volatility of the metals through the formation of hydroxides. From a thermodynamic point of view, steam could be used as an "inert" sweep gas during thermal reduction. The enhanced volatility at these temperatures may eliminate this possibility, and also raise concerns regarding steam crossover from the oxidation zone. CO_2 appears to have no effect on volatility. Initial experimental results for un-substituted-Fe/YSZ composites indicate significant mass loss occurs at temperatures of 1500 °C, a condition above the melting point of FeO and near the melting points of Fe_2O_3 and Fe_3O_4.

Conclusion

Energy is by far the largest human endeavor on the planet; global expenditures measured roughly $3 trillion in 2005, or more than twice the amount spent on agriculture and 3 times the amount on defense [29,30]. Tremendous capital is invested in the current hydrocarbon infrastructure, however massive capital investments will continue to be necessary, even if we choose to preserve the fossil fuel paradigm. Applying renewable energy resources to recycle CO_2 is a new paradigm for storing and transporting energy that preserves much of the hydrocarbon infrastructure, while offering a more environmentally sound future. However, achieving the chemical conversions with high energy efficiency is a necessity. Thermochemical cycles for splitting water and carbon dioxide have promise as an efficient approach to realize this paradigm, but the technology for implementation is immature. In our work, multi-cycle CO_2 and H_2O splitting has been demonstrated over both iron- and ceria-based monolithic materials of the type required for a solar reactor such as the CR5. Increased reaction kinetics are desirable and would decrease the required heat exchange efficiency in the CR5, for example. The long term durability of the current generation of reactive monoliths is being evaluated, even as improved structures and materials are being designed. Future work will include benchmarking the efficiency of the current generation CR5.

Acknowledgement

This work was supported by the Laboratory Directed Research and Development program at Sandia National Laboratories, in the form of a Grand Challenge project entitled "Reimagining Liquid Transportation Fuels: Sunshine to Petrol." Sandia is a multiprogram laboratory operated by Sandia Corporation, a Lockheed Martin Company, for the United States Department of Energy's National Nuclear Security Administration under Contract DE-AC04-94AL85000.

---

1. For an overview see S. Han and C. D. Chang "Fuels, Synthetic, Liquid Fuels" in *Kirk-Othmer Encyclopedia of Chemical Technology,* John Wiley & Sons, Inc. (1994).

2. Carbon Management: Implications for R&D in the Chemical Sciences and Technology, National Academies Press, 2001.

3. Carbon Dioxide Recovery and Utilization, M. Aresta, ed., Kluwer Academic Publishers, Dordrecht, 2003.

4. X. Xiaoding and J.A. Moulijn "Mitigation of CO_2 by Chemical Conversion: Plausible Chemical Reactions and Promising Products" *Energy and Fuels*, 1996, **10**, 305.

5. Chemical & Engineering News July 2, 2007

6. U.S. Energy Information Administration Report "Emissions of Greenhouse Gases in the United States 2005" DOE/EIA-0573 (2005).

7. J. S. Dukes "Burning buried sunshine: Human consumption of ancient solar energy" Climatic Change 61 (2003) 31.

8. X.-G. Zhu, S. P. Long, and D. R. Ort "What is the maximum efficiency with which photosynthesis can convert solar energy into biomass?" Current Opinion in Biotechnology 19 (2008) 153.

9. N. S. Lewis and D.l G. Nocera "Powering the planet: Chemical challenges in solar energy utilization" PNAS 103 (2006) 15729.

10. P. A. Lehman, C. E. Chamberlin, G. Pauletto and M. A. Rocheleau "Operating Experience with a Photovoltaic-Hydrogen Energy System" Int. J. Hydrogen Energy, Vol. 22, No. 5, pp. 465470, 1997.

11. C. A. Henao, C. T. Maravelias, J. E. Miller, and R. A. Kemp "Synthetic Production of Methanol Using Solar Power" in *Design for Energy and the Environment: Proceedings of the Seventh International Conference on the Foundations of Computer-Aided Process Design* Editors: M. M. El-Halwagi and A. A. Linnenger, CRC Press (2009).

12. D. Mignard, M. Sahibzada, J. M. Duthie, H.W. Whittington "Methanol synthesis from flue-gas CO_2 and renewable electricity: a feasibility study" International Journal of Hydrogen Energy 28 (2003) 455.

13. D. Mignard and C. Pritchard "Process for the synthesis of liquid fuels from CO_2 and marine energy" Trans IChemE, Part A, Chemical Engineering Research and Design, 2006, 84(A9): 828–836.

14. J. K. Stolaroff, D.W. Keith, G.V. Lowry "Carbon dioxide capture from atmospheric air using sodium hydroxide spray" Environmental Science and Technology" 42(8) (2008) 2728.

15. R. Wiser, G. Barbose, and C. Peterman "Tracking the sun: The installed cost of photovoltaics in the U.S. from 1998-2007" LBNL Report LBNL-1516E, February 27, 2009 Revision.

16. M. Green, *Annual Report, Third Generation Photovoltaics*, University of New South Wales, Sydney, Australia (2000).

17. J. E. Miller, M. D. Allendorf, R. B. Diver, L. R. Evans, N. P. Siegel, and J.N. Stuecker "Metal Oxide Composites and Structures for Ultra-High Temperature Solar Thermochemical Cycles" Journal of Materials Science 43 (2008) 4714 **DOI:** 10.1007/s10853-007-2354-7.

18. M. D. Allendorf, R. B. Diver Jr., N. P. Siegel, and J. E. Miller "Two-Step Water Splitting Using Mixed-Metal Ferrites: Thermodynamic Analysis and Characterization of Synthesized Materials" Energy & Fuels 2008, 22, 4115–4124.

19. T. Kodama and N. Gokon "Thermochemical cycles for high-temperature solar hydrogen production" *Chem. Rev.* **2007,** *107*, 4048.

20. J. E. Miller "Initial Case for Splitting Carbon Dioxide to Carbon Monoxide and Oxygen" SAND2007-8012, December 2007.

21. R. B. Diver, J. E. Miller, M. D. Allendorf, N. P. Siegel, R. E. Hogan "Solar Thermochemical Water-Splitting Ferrite-Cycle Heat Engines" Journal of Solar Energy Engineering 130 (2008) 041001-1. 19 (2008) 153.

22. J. E. Miller, L. R. Evans, J. N. Stuecker, M. D. Allendorf, N. P. Siegel, R. B. Diver "Materials Development for the CR5 Solar Thermochemical Heat Engine" Proceedings of ASME International Solar Energy Engineering Conference 2006, Denver, CO, ISEC2006-99152.

23. J. E. Miller, L. R. Evans, N. P. Siegel, R. B. Diver, F. Gelbard, A. Ambrosini, and M. D. Allendorf "Summary Report: Direct Approaches for Recycling Carbon Dioxide into Synthetic Fuel" SAND 2009-0399, January 2009.

24. J. N. Stuecker, J. Cesarano III, and D. A. Hirschfeld, Journal of Materials Processing Technology 142 (2003) 318.

25. N. P. Siegel, R. B. Diver, J. E. Miller, T. J. Garino, S. Livers "Reactive Structures for Two-Step Thermochemical Cycles Based on Non-Volatile Metal Oxides" Proceedings of ES2009 Energy Sustainability 2009 July 19-23, 2009, San Francisco, CA USA.

26. E. N. Coker, M. A. Rodriguez, A. Ambrosini, R. R. Stumpf, E. B. Stechel, C. Wolverton and B. Meredig "Fundamental Materials Issues for Thermochemical H_2O and CO_2 Splitting: Final Report (FY08)" Sandia Report SAND2008-7655, November 2008.

27. D.L. James, N.P. Siegel, R.B. Diver, B.D. Boughton, and R.E. Hogan "Numerical Modeling of Solar Thermo-Chemical Water-Splitting Reactor" Proceedings of ASME International Solar Energy Engineering Conference 2006, Denver, CO, ISEC2006-99141.

28. K.S. Chen and R.E. Hogan "A Two-Phase Model for Solar Thermochemical Water Splitting with FeO/Fe_3O_4" Proceedings of ES2009 Energy Sustainability 2009 July 19-23, 2009, San Francisco, California USA ES2009-90228.

29. Richard Smalley "Nanotechnology and Our Energy Challenge" in Nanotechnology: Science, Innovation, and Opportunity, Lynn E. Foster, ed., Prentice Hall PTR, 2005.

30. Richard E. Smalley "Future Global Energy Prosperity: The Terawatt Challenge" MRS Bulletin 30(6) 412.

Energy Technology 2010
Edited by: Neale R. Neelameggham, Ramana G. Reddy, Cynthia K. Belt, Ann M. Hagni and Subodh Das
TMS (The Minerals, Metals & Materials Society), 2010

Materials Engineering Aspects of Intermetallic Anode for Aluminium Electro-Winning

Animesh Jha[1], Greg Doughty[2] and Xiaobing Yang
[1]The Institute for Materials Research, Houldsworth Building, University of Leeds, Leeds LS2 9JT, UK
[2]Inertius Ltd, Innovation Technology Centre, Advance Manufacturing Park, Brunel Way, Rotherham, S Yorkshire S60 5WG

Successful use of inert anodes should reduce the CO2 footprint in aluminum production. The search for a "holy grail" inert anode for metal electro-winning from constituent oxides using molten salt bath as an electrolytic medium is continuing. In this article we demonstrate the engineering approaches for the fabrication of intermetallic anodes, based on a quaternary Al-Ti-Cu-M (M:Ni, Fe) system for primary aluminium production. In particular, we examine the microstructural and electro-chemical properties of alloys, based on the quaternary compositions. We explain the aspects of phase stability under anodic potential, which leads to the loss of anodic materials in the temperature range of 850-1050°C. The results are analysed in the context of oxidation reaction and formation of oxide layers which readily dissolve in the cryolite melt and prevent passivation. The paper also describes the conditions under which the reduction of aluminium at the graphite cathode occurs which also strongly depends on the molten salt chemistry for the nucleation of molten aluminium at the cathode.

Keywords: anode, cryolite, corrosion, intermetallic phase, aluminium

1. Introduction

Oxygen-evolving inert anodes are an alternative to the conventional consumable carbon anodes for the production of primary aluminium. In this paper the materials selection criteria chosen for the engineering of anode is based on the understanding of the thermodynamic stabilities of aluminide intermetallic phases, some of which are formed and remained stable in fluoride melts. For example, the formation Al_3Ti dispersed in aluminium metal via molten-fluoride flux reaction is well known in the context of aluminium casting and grain refining [1-3]. It is for this reason in our research we concentrated on the engineering of stable intermetallic phase mixture in the Al-Ti-Cu ternary system, shown in Figure 1 [4]. In this figure two compositions in the $Al_{3-0.5z}Ti_{1-0.5z}Cu_z$ family are shown with the substitution of 7 atomic % and 11 atomic % copper. However initial investigations showed that these two compositions were unsuitable due to the formation of low-temperature liquid (~540°C) and intermetallic phase, Al_2Cu along the Al-Cu binary edge during solidification. The Al_2Cu phase segregated along the boundaries of Al_3Ti and Al_5Ti_2Cu, and dissolved away when electrolysis was carried out using these anodes above 800°C. The presence of consequential porosity in the anode material increased the corrosion of the alloy through local heating due to high current density at 0.8 A cm^{-2}. The early tests were carried out with cryolite melts, shown in Figure 2, having 3 weight percent of Al_2O_3 in the bath at 850°C.

In this paper we investigate the most stable, and therefore, the dominant phase field in the ternary alloy composition, shown in Figure 1. From the analysis of phases, it is apparent that the corrosion resistance compositions may be those, which exhibit equilibrium high-temperature intermetallic phases, Al_5Ti_2Cu and $AlTi_2Cu$. Although initially we focussed on 7 at% Cu and 11 at% Cu concentrations, the knowledge on corrosion and low-melting point phases and reactions helped us in developing the new multi-component compositions, in which the concentrations of aluminium was reduced and that of titanium increased, as shown in Table 1. Since the ternary alloy analysis with compositions in excess of 5 at% Cu showed the formation of extensive grain boundary, Al_2Cu, phase, implying that the excess copper cannot remain dissolved in the matrix dominated by Al_5Ti_2Cu phase above 800°C. For this reason we considered the incorporation of Ni, which alloys with copper well and forms extensive solid and liquid solution in the binary Ni-Cu system. In order to reduce the segregation of Al_2Cu, metallic Fe was incorporated in the matrix, as shown in Table 1, for lowering the chemical potential of Al by forming strong intermetallic phases, namely Al_3Fe and Al_3Ni which then dissolve in the ternary intermetallic.

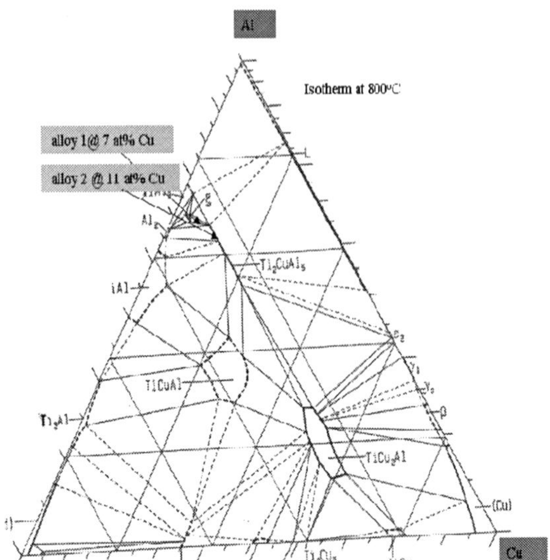

Figure 1: An isothermal section of Al-Ti-Cu ternary section at 800°C, showing two alloy compositions (alloys 1 and 2) with 7 and 11 atomic percents (at %) of Cu in a composition $Al_{3-0.5z}M_{1-0.5z}Cu_z$ [4]. The ring shown closer to Al-Ti edge is the region of preferred composition for anode engineering

The oxidation resistance of alloy was studied by thermogravimetry, scanning electron microscopy, and oxidation studies using molten cryolite. Once the oxidation analysis on new alloy compositions were completed, the anodes were prepared for cell tests between 6 hours and 24 hours. Longer than 24 hrs cell tests are essential for determining the longevity of anodes in the service environment. The anticipated corrosion rate of anode, which industry expects, for using them in Al-cell is less than 20 mm per year.

Table I: Compositions of multi-component alloy based on Al-Ti-Cu ternary, used for anode engineering, which were cast and annealed for anodic oxidation

Composition (atomic %)	Alloy 1	Alloy 2	Alloy 3
Al	75- (z/2)	68	45
Ti	25-(z/2)	23	40
Cu	z	3	5
Ni	0	2	5
Fe	0	2	5
Cr	0	2	0

2. Experimental

The compositions of three alloys, shown in Table I, were prepared by the conventional high-temperature melting in a controlled atmosphere above 1585°C for 3 hours under the shroud of Ar-gas. The homogenised liquid was then cast into a preheated steel mould at 550°C by pouring into the mould gently by minimising the trapped gases, so that the cast surface remained smooth and shiny. The mould with alloy was then placed back into the annealing furnace for few hours at 550°C, after which the furnace power was turned off allowing the cast alloy to cool slowly. The high-temperature oxidation behaviour of these alloys was studied between 850°C and 1050°C in air by heating the sample in a thermogravimetric balance. The airflow rate in the balance was 0.5 l min^{-1}. After each isothermal run the alloy sample was prepared for scanning electron microscopic examination, which is discussed below. Besides dry oxidation analysis in air, the oxidation reactions for alloy samples were studied in a cryolite bath by choosing the composition shown in Figure 2. Each salt-bath corrosion test lasted for 4 hours in air at the chosen temperature of 900°C and 950°C. For alumina reduction, the cryolite was then mixed with 3 wt% alumina. For the analysis of passivating phase present on the surface of anode, we adopted two-electrode cyclic voltammetry, which has been proposed by Chen and co-workers [5]. The cyclic voltammetry tests were carried out at 950°C using an Autolab Potentiostat PGSTAT10 module. In the cell the Al-Ti-Cu alloy was the working electrode, whereas the graphite rod was used as the counter electrode. Both alloy and carbon electrodes were 13 mm in diameter and immersed in the cryolite bath to a depth of 1 cm for maintaining a uniform current density on each electrode. The microstructure of each alloy was analyzed by using a scanning electron microscopy (Cambridge Camscan 4 FEG SEM) operated at 20 kV. Chemical analysis of the phases present was then carried out using the "Link Oxford" energy dispersive X-ray spectroscopy (EDS) system on the scanning electron microscopy.

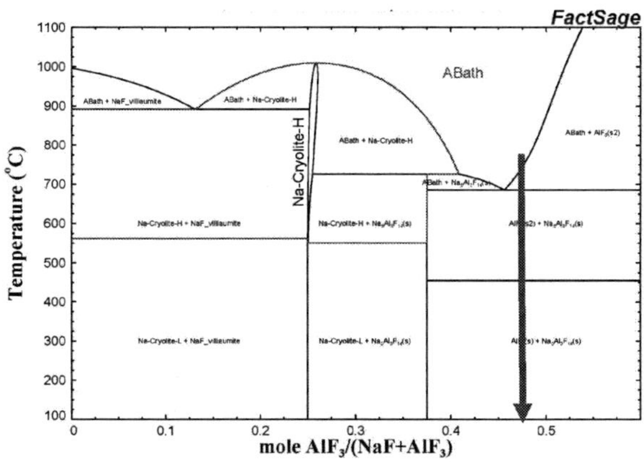

Figure 2: Sodium cryolite phase diagram used for the dissolution of alumina with 3 wt% of alumina for aluminium electrolysis [5].

3. Results and discussion

3.1) Preparation of Alloy Compositions and characterisation

The X-ray diffraction of phases formed in the cast and annealed alloy compositions, summarised in Table I, are compared in Figure 3 in which we observe that the main reflection at around 39° Bragg angle for Al_5Ti_2Cu shifts as shown in the inset and is dependent upon the composition suggesting that the transition metals may be replacing both Ti and Cu in the structure of the intermetallic. Except in Alloy 3, the $Al_3(Ti,M)$ was the dominant phase in Alloy 1 and 2. The microstructure of Alloy 1 is shown in Figure 4. There is clear evidence that the Al_3Ti rods are fragmenting, indicating that if the Al_3Ti phase forms in the matrix of Al_5Ti_2Cu, it tends to decompose via dissolution, which is apparent from the phase equilibrium condition, derived from the ternary section in Figure 1. With the additions of transitional metal Ni, Fe and Cr, the relative abundance of light grey $Al_5(Ti,M)_2(Cu, Ni)$ and bright $Al_2(Cu,M)$ phases increase in Figure 5 with respect to $Al_3(Ti,M)$ (dark phase) in the composition, $Al_{5-0.5z}(Ti,M)_{2-0.5}Cu_z$, in alloy 1 as shown in Figure 4. The corresponding energy dispersive spectra (EDS) for three identified phases (dark, ligh grey, and bright) are compared. With the increasing concentrations of Ti, as shown Figure 6 for alloy 3, only two phases are present: the dark phase AlTi and the light colour phase is a solid-solution of AlTi with transition metals. The EDS analysis reveals that almost all the transition metals dissolved in alloy 3 were distributed in the light phase. By comparison, in alloy 2 the transition metals are mainly present in the grey and bright phases. The analytical SEM also showed that the small concentrations of Cr in the microstructure of Al_3Ti phase.

As a part of the alloy fabrication, we also analysed the effect of slow cooling, say of alloy 2 shown in Figure 7, as compared to a structure obtained via fast cooling and subsequent annealing shown in Figure 5. Extensive grain boundary segregation, as shown in Figure 7, occurs on slow cooling as it is apparent from the presence of spherical shape phases at the grain boundaries, indicating a presence of the last liquid during slow solidification.

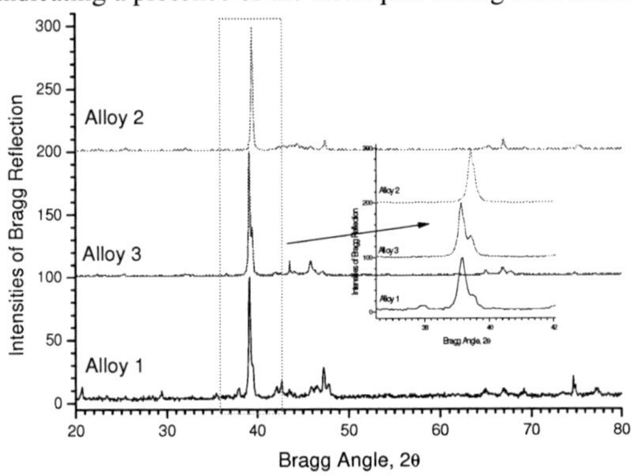

Figure 3: A comparison of X-ray diffraction data (Cu K_α) for three alloy compositions studied. The main peak for Al_5Ti_2Cu phase shifts with the composition of alloy.

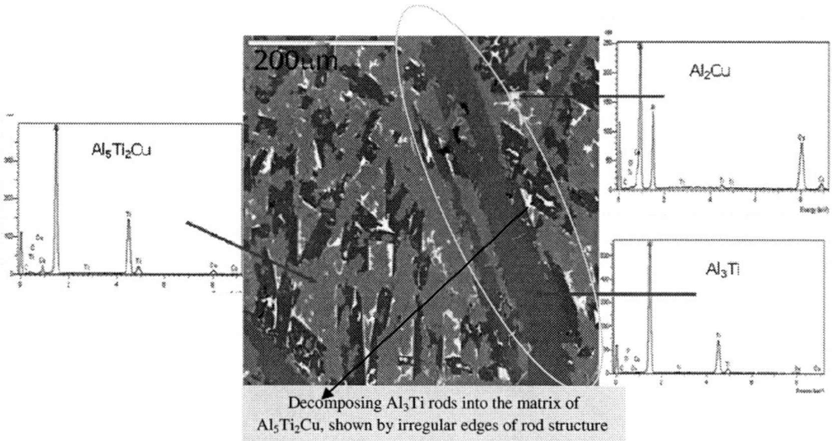

Figure 4: The microstructure of alloy 1, showing 3 different intermetallic phases, as identified by EDS. The Al_3Ti (dark grey) phase in the centre of the microstructure is decomposing and dissolving into the matrix dominated by Al_5Ti_2Cu matrix, in which we also see the presence of Al_2Cu.

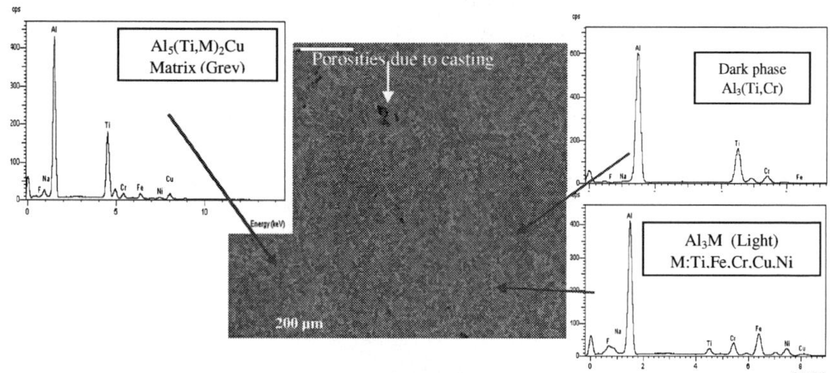

Figure 5: Microstructure as-cast alloy 2 with its constituent phases, shown in light Al_3M (M: Cu, Ti, Fe, Ni, Cr), grey $Al_5(Ti,M)_2Cu$ and dark grey $Al_3(Ti,Cr)$ shades of colours.

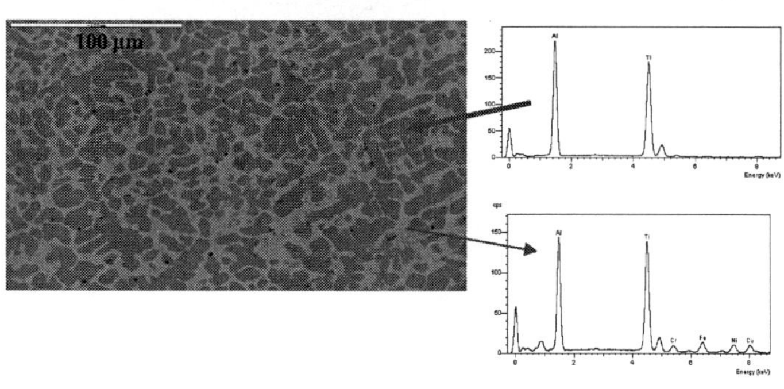

Figure 6: Microstructure of alloy 3 with two dominant phases, AlTi and $Al_5(Ti,M)_2Cu$

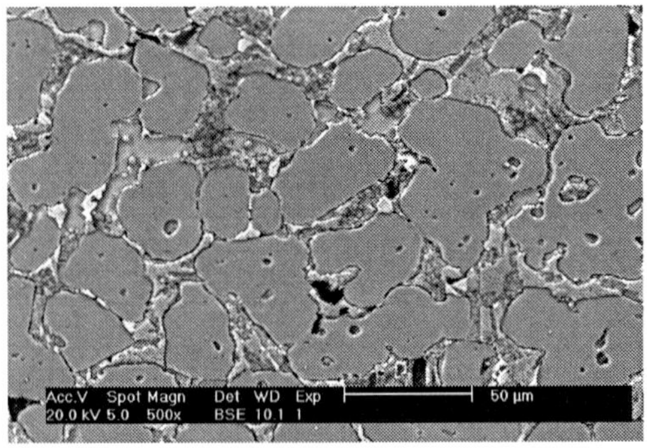

Figure 7: The microstructure of the as-cast alloy 2 under slow cooling, showing extensive grain boundary segregation and micro-porosity

3.2) Oxidation tests in air and with molten cryolite in air

The high-temperature oxidation tests were carried out using thermal gravimetric technique in a stream of air between 850°C and 1050°C, using isothermal and isochronal scans, as shown in Figure 8. As expected, the oxidation rate of alloy increases with temperature and there appears to be a process of reaction slowing down and levelling off at all temperatures. Significant of oxides of aluminium with other transition metal oxides are observed on the surface, as shown in Figure 9.

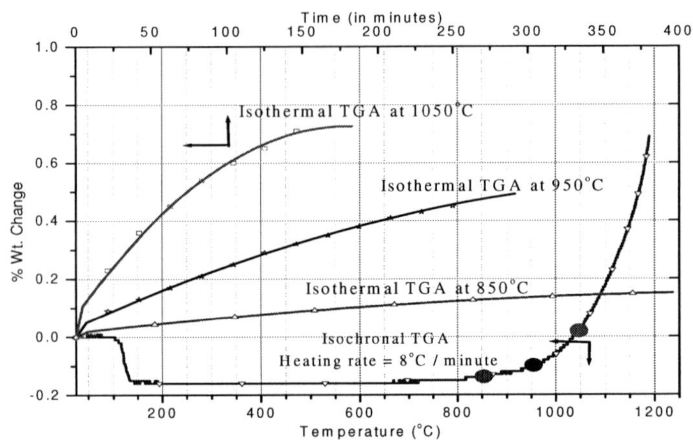

Figure 8: Thermogravimetric analysis of alloy 1 between 850°C and 1050°C under isochronal and isothermal conditions in an atmosphere of air, maintained at 0.5 l min^{-1}.

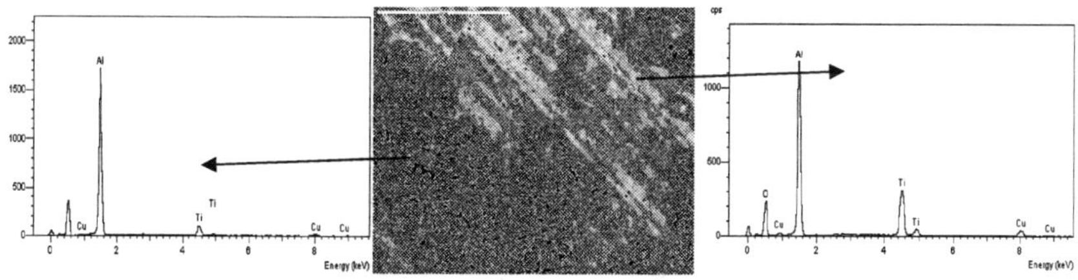

Figure 9: SEM image accompanied with the oxidation product analysis on the surface of alloy 1, oxidised at 950°C for 4 hours in air. Two types of oxide layers are apparent from the EDS analysis.

The two-phase oxide constitution on the surface of alloys depicts the preponderance of Al_3Ti and Al_5Ti_2Cu phases in alloy 1. The relative intensities of X-ray peaks in the EDS also confirm the correspondence of oxides with the parent alloy phases. The oxidation behaviour of alloy 2 and alloy 3 were much slower than alloy 1 at 850°C as shown in Figure 8. Based on the comparison of oxidation behaviour of alloys 1-3, in the next phase of oxidation and corrosion analysis the focus was to analyse the performance of alloys 2 for salt base corrosion in air at and above 850°C.

For high temperature oxidation tests for alloy 2 in molten cryolite, the composition of salt bath selected was based on the data in Figure 2. The selected composition was close to the eutectic composition in Figure 2. There is clear evidence from the analytical SEM data in Figures 10a to 10c that the alloy corrodes by preferentially loosing aluminium as oxide into the molten cryolite. Consequently the interfacial region loses aluminium, leaving behind a copper-nickel-iron rich layer, which is preferentially present in a titanium rich matrix. The line scans in 10c also confirms clearly that there is more Al on the surface than in the bulk and it is varying in a sinusoidal manner, compared to Cu which is concentrated behind the Al-layer at the surface.

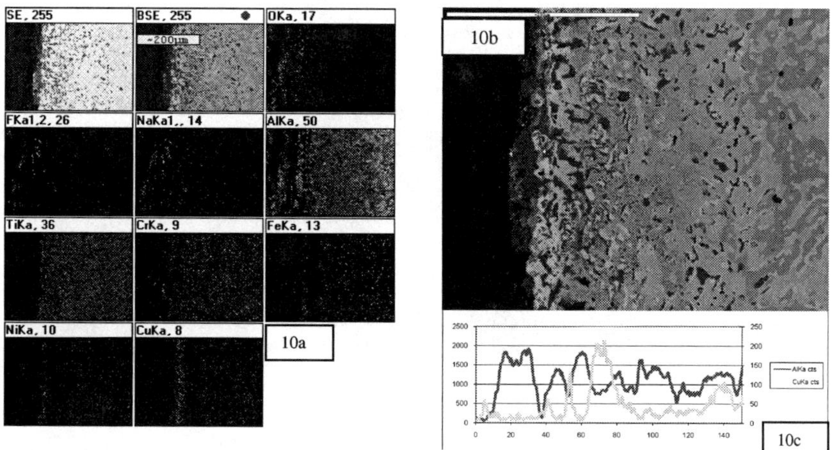

Figure 10: The microstructure analysis of alloy 2 after oxidation in molten cryolite mixture at 900°C, shown by arrow in Figure 2. 10a) shows elemental analysis; 10b) is backscattered image; and 10c) compares two line scans for each Al (pink) and Cu (yellow). From the analysis it is apparent that aluminium is preferentially lost during oxidation/corrosion reaction from the surface of alloy 2.

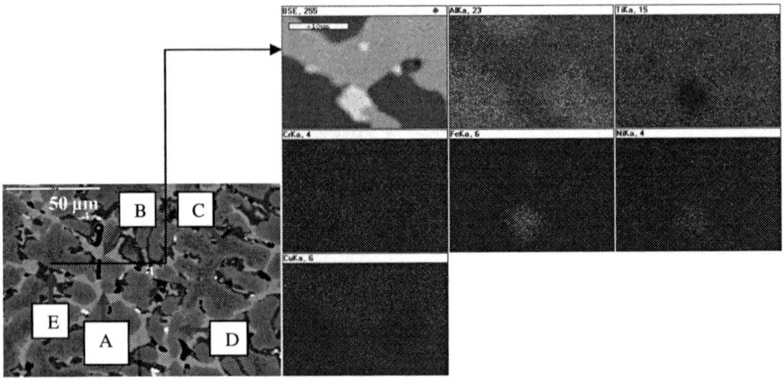

Figure 11: A high-magnification backscattered SEM image of alloy 2, after interfacial oxidative corrosion in molten cryolite at 900°C in air. The accompanying elemental map represents the Al-rich area E in the figure identified by black arrow. Area A: transition metal rich Ti-Al phase; Area B: is the micro porosity, Area C: is the decomposing Al_3Ti phase, Area D: predominantly rich in Fe/Ni, and Area E: in aluminium.

In order to analyse the loss of aluminium from alloy in detail we have examined the microstructure at higher magnification, from which we find that the Al_3Ti phase (dark phase area C in Figure 11) is rapidly decomposing and disappearing into the matrix, leaving micro-porosity, B and Ti, Fe, Cu, and Ni behind (area A). As a first approximation in phase decomposition analysis, it is reasonable to consider that Al_3Ti is unstable under the oxidizing condition, leading to formation of alumina which dissolves in the molten cryolite. In the oxidised alloy there is also evidence for the ever-reducing transition metal (Fe,Ni) rich phase (D), as shown in Figure 11a which becomes apparent when the EDS image of a representative area is compared for elemental distribution.

Typical cyclic voltammograms were obtained using two-electrode technique in the presence and in the absence alumina in cryolite, and these are compared in Figures 12a to 12d. For comparative purposes, the carbon working electrode was also used in the test rig. A careful examination of these cyclic voltammograms leads to the following conclusions. First, during cathodic scan, no deposition of materials was found for both alloy and carbon working electrodes in the absence of alumina in cryolite at 950°C (see Figure 12a). When the scan was reversed, no anodic peak was observed in the absence of alumina (see Figure 12b). When 3 wt% alumina was mixed in the cryolite bath, the cathodic peak was observed immediately with different working electrodes, indicating the reduction reaction of alumina was occurring in case of carbon and alloy. During anodic scan, only one anodic peak was found in alloy electrode (Figure 12 c) and three anodic peaks for carbon anode (Figure 12d). As we were unable to use any reference electrode, it was impossible to analyse the mechanism of oxidation reaction conclusively. Nevertheless, a qualitative comparison between the alloy and carbon working electrodes indicates that the alloy 2 anode is more stable than carbon anode in a cryolite bath.

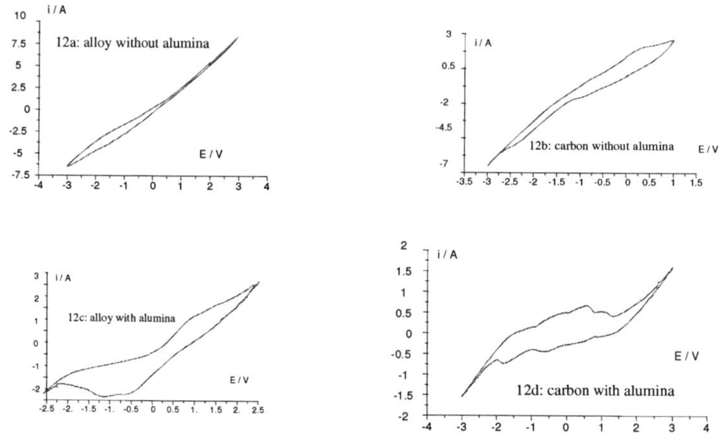

Figure 12: A comparison of cyclic voltammetry scans for cryolite melts and carbon and alloy 2 as working anodes at 950°C in the absence and the presence of 3 wt% alumina in cryolite. The peaks in 12c (alloy) and 12d (carbon) are indicative of reactions occurring in the cathodic and anodic cycles. Scan rate was carried out 0.2V.

The main conclusions from oxidation analysis are that the alloys with lower concentrations of Ti and transition metals tend to oxidise, preferentially via the loss of aluminium metal which forms an oxide layer on the surface. Since alumina formed rapidly dissolves in cryolite, the oxidation reaction continues until either the diffusion of aluminium in alloy becomes a barrier, or all the aluminium from the alloy is lost. The tendency for the loss of aluminium in alloys 2 and 3 having much higher concentrations of transition metals than in alloy 1 reduces rapidly, which on the basis of thermodynamic equilibrium points out that as the chemical potential of aluminium reduces in the alloy, the oxidation under anodic potential becomes less and less significant and therefore contributes to long term anode stability in cryolite melt below 900°C.

3.3) Anode Reaction in Cell:

We melted 200 ml of cryolite with a composition shown in Figure 2 and maintained the bath temperature at 850°C by heating it inside a resistive furnace. The cell with carbon cathode and inert anode was purged with nitrogen gas and the molten cryolite at 850°C was then treated with 3 wt% alumina. The anode-cathode pair was then lowered and 4 Amperes of current was passed through the two electrodes. Since the crucible was only 100 mm in diameter the anode-cathode distance was less than 50 mm. Approximately 1 cm of anode was lowered into the bath and the electrolysis was started. Time-to-time alumina was fed by mixing with cryolite of composition, shown in Figure 2. Experiments were carried out from few hours to 24 hours. Major problem in running longer period experiment was in maintaining the volume of cryolite, which often led to voltage drop, as shown in Figure 13a. The reduction was carried out at 850°C. The lack of stirring due to natural convection of cryolite was the primary reason for the discontinuity in the voltage-time diagram. The voltage drop was due to the crust formation in a small cell, which raised back to 4.8V in Figure 13b, when the anode-cathode pair were shaken after an overnight standstill. At the end of the run, the crucible was broken deliberately at the top of frozen salt line for examination of the melt surface and any potential proof for alumina reduction. In Figure 13b we find the droplets of aluminium which formed during the two uninterrupted periods in the cell 0-20,000 sec and 66000-91000 sec. The aluminium metal and cryolite bath were analysed for trace impurities. From EDX and lattice parameter analyses of aluminium little solutes were found in the pure aluminium matrix. The close examination of anode did show some evidence for local anode erosion around the salt line otherwise it was found to be in a satisfactory state. The metal microstructure and EDX analysis also did not show any evidence for the presence of impurities. Occasionally silicon as an impurity in the matrix was found which might have originated from carbon cathode. We weighed the aluminium found which was 9.6 g and corresponded to 85% current efficiency and is much lower than in the conventional cell. However the short-fall in the current efficiency is evident from a long period of interruption in the voltage-time diagram in Figure 13a. Figure 13c shows the photo-micrograph of anode with minor surface erosion, which may have arisen during the electrical discontinuity in the bath.

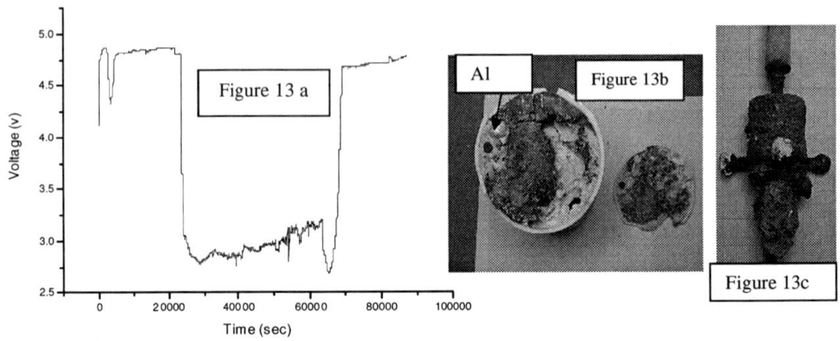

Figure 13: a) Voltage-time plot for reduction of alumina using alloy 2 at 850°C. b) Photo-micrograph of aluminium produced underneath the cathode near the hole, where the cathode was positioned. c) shows the anode which was covered in frozen salt.

4. Conclusions

The Al-Ti-Cu based intermetallic inert anodes exhibit good oxidation resistance in air and cryolite salt bath at and below 850°C. The Al_3Ti is the main phase which appears to be unstable in the salt-bath under oxidising condition and it decomposes and forms the intermetallic rich in transition metals, as shown in microstructures in Figures 4, 9, and 11. The presence of transition metals, such as nickel and iron in the alloy 2 composition, stabilises the material against oxidation. Limited numbers of cyclic voltammetry tests also suggest that the oxidation reaction with alloy is quite small in magnitude, compared with the carbon as an anode material. On the basis of extensive oxidation analysis of alloy compositions, we undertook extensive cell test for aluminium production in the 700 to 850°C and proved that alumina was reduced to aluminium metal, as shown in Figure 13 over a period of 24 hours. The 24-hour cell reaction shown in this figure was disrupted as a result of alumina crust formation. When the electrodes were shaken gently the electrolysis recommenced, as has been shown in Figure 13a. From the Faraday's law the cell current efficiency at 850°C was ~85%.

Acknowledgements

The authors acknowledge financial support from the Engineering and Physical Science Research Council, UK between 2005 and 2008 and the Inertius Ltd.

References

1. D G McCartney (1989) Grain refining of aluminium and its alloys using inoculants. *Int. Mater. Reviews*, **34** (5), 247–260.
2. P Schumacher and A L Greer. (1996) High-resolution transmission electron microscopy of grain-refining particles in amorphous, aluminium alloys. *Light Metals* (ed. by W. Hale), The Minerals, Metals & Materials Society, Warrendale, pp. 745–753.
3. X. Wang, A. Jha, R. Brydson, J. Ellis (1999), Microstructural analysis of Al alloys dispersed with TiB_2 particulate for MMC applications, J Microscopy 196 (2) (1999) 137–145.
4. "Ternary alloys" (1988) Eds. G. Petzow and G. Effenberg, Publisher, Weinheim : VCH, vol. 8, 1988, 51 – 67.
5. L C Chen and K C Ho (2001), Electrochimica Acta, Vol.46, 2159-2166.
6. C. Bale, A.D. Pelton, W.T. Thompson, J. Melancon, and G. Eriksson: *FACT-Sage 5: Facility for the Analysis of Chemical Thermodynamics*, Ver. 7.0, CRCT, École Polytechnique de Montréal, Montréal, 2009.

Energy Technology 2010
Edited by: Neale R. Neelameggham, Ramana G. Reddy, Cynthia K. Belt, Ann M. Hagni and Subodh Das
TMS (The Minerals, Metals & Materials Society), 2010

SYNTHESIS AND CARBON DIOXIDE ADSORPTION OF ZEOLITIC IMIDAZOLATE FRAMEWORKS

Jinghua Zou[1], Huimin Lu[1*], Min Li[1]

[1]Beihang University, School of Mater. Sci. & Eng.;
37 Xueyuan Road, Haidian District, Beijing 100191, China
[*] Corresponding author: lhm0862002@yahoo.com.cn

Keywords: carbon dioxide adsorption, porous materials, solvothermal synthesis, zeolitic imidazolate frameworks (ZIFs)

Abstract

Zeolitic imidazolate frameworks (ZIFs) are new type of porous materials, form inerratic polyhedral in three-dimensional with large cages and small apertures, and appear structural diversity. Their hole size can be adjusted. These materials have high selective adsorption capacity to carbon dioxide. In this paper, zeolitic imidazolate frameworks were synthesized by solvothermal synthesis method with dimethylformamide (DMF) as solvent and structure-directing agent, divalent metallic ions as cation, and imidazole as ligand. The crystals were characterized with XRD, SEM for their crystal structures and physical properties, the adsorption of the crystals for carbon dioxide was also tested, and the result suggests that these porous materials have a capacity of 1.35ml/g for storing carbon dioxide at 273 Kelvin under ambient pressure.

Introduction

Global warming caused by the increase of carbon dioxide in the atmosphere has become one of the main environment problems, contemporary research has focused on ways to slow or stop this trend. CO_2 capture has attracted attention due to the potential to trap and sequester large amounts of CO_2 from concentrated sources such as power plants [1]. Traditional technologies in capturing CO_2 include absorption by aqueous amines [2]. However, utilization of this process is energy intensive and expensive. Carbon dioxide storage materials investigated so far can be divided into two categories: chemical storage adsorbents and physical storage media [3]. The first kind of materials can chemically adsorb carbon dioxide within their structures and need thermal energy to release the adsorbed carbon dioxide. The examples of the chemical adsorbents include aqueous amines, gas-hyperbranched aminosilicas [4]. The physical adsorbents store carbon dioxide by physical adsorption of carbon dioxide at elevated pressure and release the adsorbed carbon dioxide at low pressure. The physical adsorbents that have been investigated for carbon dioxide storage include zeolites, carbonaceous materials and other porous materials. Until recently, a new type of microporous material, known as zeolitic imidazolate frameworks (ZIFs), has been reported as promising adsorbents for carbon dioxide storage [5 ~ 10].

Zeolitic imidazolate frameworks, also known as coordination polymers, are ideal crystalline substances for gas separation and storage, and form inerratic polyhedral in three-dimensional with large cages and small apertures [11]. Their holes depend on the inherent structural features of zeolitic primary building unit (PBU), namely the corner-sharing tetrahedral TO_4 (T = Si or Al, O-T-O = 109.5°, T-O-T = 145° in which the T-O bonds could freely rotate) [12, 13]. It is this type of PBU that endows the zeolites with the flexibility of T-T-T angles and also make possible the various secondary building units (SBUs) and coplanar T-atom rings with 4- to 10-members. Thus, the creation of ZIFs with augmented zeolitic structures is not due to the augmengtation of a specialized zeolitic structure, but rather to the augmentation of the PBU of zeolites. Cobalt(Ⅱ) and zinc(Ⅱ) imidazolate polymers chosen as the ideal metal-organic analogues of inorganic silicates as their PBU, [M(im)$_4$] (M=Co^{2+}, Zn^{2+}), is structurally similar to the TO_4 of zeolites [14, 15]. Most of zeolitic topologies can be fabricated with metal imidazolates if effective synthetic strategies are employed.

In this paper, zeolitic imidazolate frameworks were synthesized by solvothermal synthesis method with dimethylformamide (DMF) as solvent and structure-directing agent, cobalt (Ⅱ) and zinc (Ⅱ) metallic ions as cation, and imidazole as ligand. The crystals were characterized with XRD, SEM for their crystal structures and physical properties, the adsorption of the crystals for carbon dioxide was also tested.

Experimental

Raw materials

$Zn(NO_3)_2{\cdot}6H_2O$(AR), $Co(NO_3)_2{\cdot}6H_2O$(AR), imidazole(AR), dimethylformamide(DMF, AR), dimethylacetamide(DMA, AR), all chemicals and solvents used in the syntheses were from commercial sources and used without further purification.

Synthesis

Based on the synthesis method of literature, the solvothermal synthesis method was chosen for the synthesis of zeolitic imidazolate frameworks in this paper. The synthetic procedure was as follows: $Zn(NO_3)_2{\cdot}6H_2O$ and imidazole dissolved in DMF under room temperature with stirring respectively, then mixed together. The mixture solution was placed in a high pressure reactor. The high pressure reactor was sealed and placed in a vacuum drying oven, then heated at 373K ~ 413K for 24h ~ 100h. Finally the white products were collected by filtration and washed with DMF and absolute ethyl alcohol, then dried at 373K at vacuum condition. The same synthesis was repeated by replacing $Zn(NO_3)_2{\cdot}6H_2O$ with $Co(NO_3)_2{\cdot}6H_2O$. The ratio of materials was taken by mol ratio in all synthesis experiments.

Characterization

The crystal structure of the samples were determined from X-ray powder diffraction patterns collected on Rigaku D/max-2000 diffractometer with Cu Kα1 radiation and operated at 40KV and 40mA. The relative crystallinity was calculated by comparing the diffraction intensities of the major peaks. The crystal morphologies were observed on VEGA\\LMU scanning electron microscope. CO_2 adsorption data were measured on NOVE-2200e gas adsorption apparatus at 273K.

Results and Discussion

Analysis of synthesis conditions

$Co(NO_3)_2 \cdot 6H_2O$ used as metallic cation, the Zn^{2+}/imidazole mol ratio set from 1:1 to 1:8, let the mixture react for 24h~72h at specific temperatures, most of the products were powder crystals. The crystal structure would appear unless the imidazole mol ratio was enough; when the metallic cation replaced $Co(NO_3)_2 \cdot 6H_2O$ with $Zn(NO_3)_2 \cdot 6H_2O$, the products had high crystallinity at different mol ratios, proper crystallization times and tempertatures. $Co(NO_3)_2 \cdot 6H_2O$ and $Zn(NO_3)_2 \cdot 6H_2O$ were taken as metallic cation, M^{2+}/imidazole mol ratio was 1:8, reaction time 72h, the XRD patterns of the products were showed in Fig.1. The peak positions were similar to reported XRD patterns of ZIF-61. The crystallinity of the products synthesized by $Co(NO_3)_2 \cdot 6H_2O$ was lower than by $Zn(NO_3)_2 \cdot 6H_2O$.

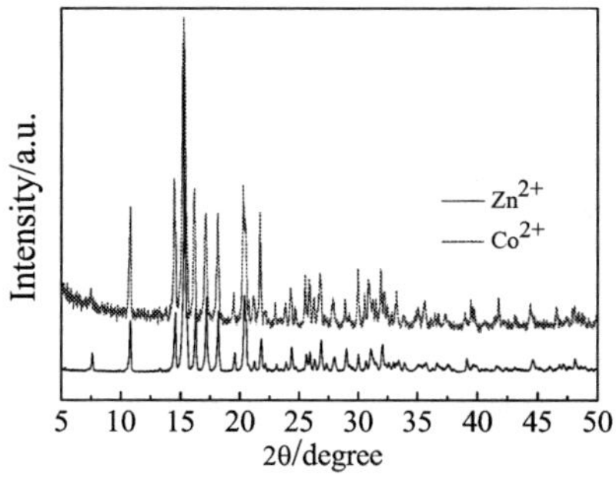

Fig.1 XRD patterns of samples with different metallic cation

$Co(NO_3)_2 \cdot 6H_2O$ used as metallic cation, there was no product at the react conditions when the temperature was 403K, react time 48h, and the mol ratio of $Co(NO_3)_2 \cdot 6H_2O$ to imidazole 1:2. When the mol ratio changed to 1:4, there were few products, but the XRD patterns showed that the products were no crystallinity. Products with the crystalline structure appeared at the mol ratio of 1:8. The result showed that the higher the mol ratio, the more crystalline the products. But $Zn(NO_3)_2 \cdot 6H_2O$ used as metallic cation and the mol ratios of $Zn(NO_3)_2 \cdot 6H_2O$ to imidazole were 1:2, 1:4 and 1:8 respectively, all products synthesized were crystal, and of the same crystalline structure. The XRD patterns were showed in Fig.2. The results suggested that increasing the molefraction of imidazole, there were more products synthesized with crystalline structure when other conditions did not change.

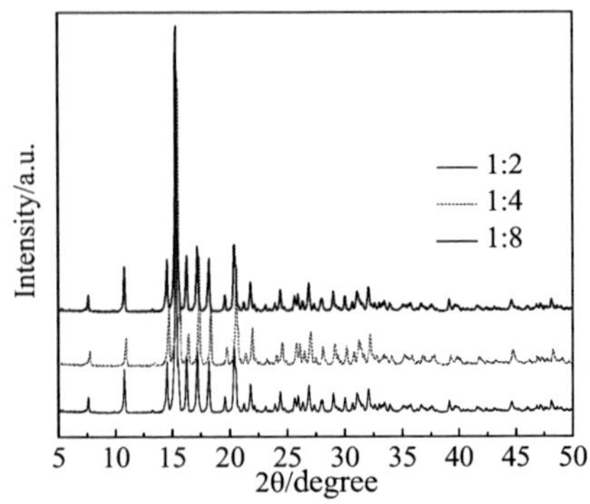

Fig.2 XRD patterns of samples with different mol ratios

It was found that the great effect of reaction temperature on the products. A series of temperatures was set for comparing the effect of different temperatures on the products. When the reaction time was 72h, the mol ratio of $Zn(NO_3)_2{\cdot}6H_2O$ to imidazole was 1:4, the temperature changed step by step. The results indicated that when the temperature was lower than 373K, there was no product synthesized. As the temperature increased, products appeared, and the crystallinity was rising. But when the temperature changed to 403K or even more higher, the crystallinity did not rise any more. The result of the effect of temperature was showed in Fig.3.

Reaction time was also a very important factor in the products. The effect of reaction time on the crystallinity of products was researched when the reaction temperature was 403K and the mol ratio of $Zn(NO_3)_2{\cdot}6H_2O$ to imidazole was 1:4. The results were showed in Fig.4. This suggested that the crystals appeared when the reaction time reached 12h, but the crystallinity was low. As the reaction time extended, the crystallinity became higher. When the reaction time reached 72h, the crystallinity did not change any longer. This follows the liquid phase reaction mechanism that the procedure needed long time to react completely.

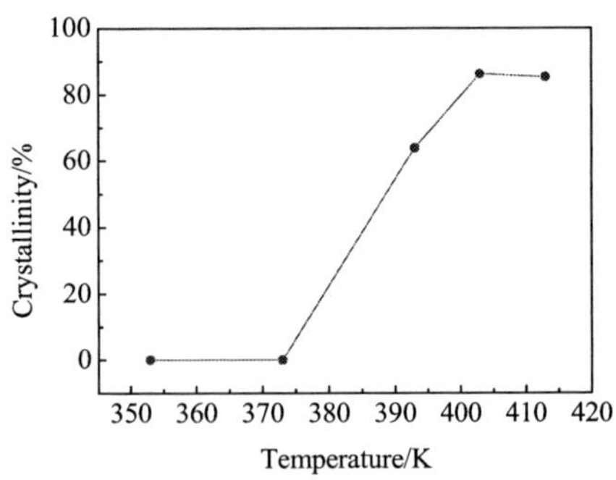

Fig.3 The effect of temperature on crystallinity

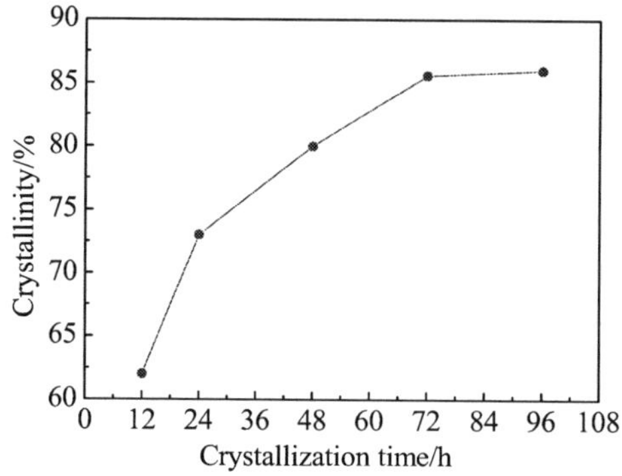

Fig.4 The effect of crystallization time on crystallinity

Morphologies of samples

The morphologies of the typical samples were observed by SEM. As showed in Fig.5, the samples synthesized were in the form of single prism-shaped crystals. The length was about 300μm ~ 600μm, the diameter was 30μm ~ 100μm. The solvothermal synthesis method could get integral crystal because they follow the liquid phase reaction mechanism. And the morphology of the single prism-shaped crystal structure matched the XRD patterns well; all these indicated that the samples were regular particles.

Fig.5 SEM of samples at 403K for 72h

Carbon dioxide adsorption measures

The carbon dioxide adsorption data were analyzed based on the carbon dioxide adsorption isotherms at 273K by NOVE-2200e gas adsorption apparatus. The samples researched were synthesized under the best condition. The organic guests contained in ZIFs would occupy the majority of void areas of framework. Generally, the organic guests may be removed away by heating or solvent extraction. In our experiments, these guests of samples were removed by heating and chloroform solvent, and the carbon dioxide storage ability was studied. The corresponding data were showed in Fig.6. The results suggested that these porous materials have capacity for storing carbon dioxide: 1.35ml/g at 273 Kelvin under ambient pressure. And the carbon dioxide storage of the sample synthesized for 72h was higher than that of the sample synthesized for 48h.

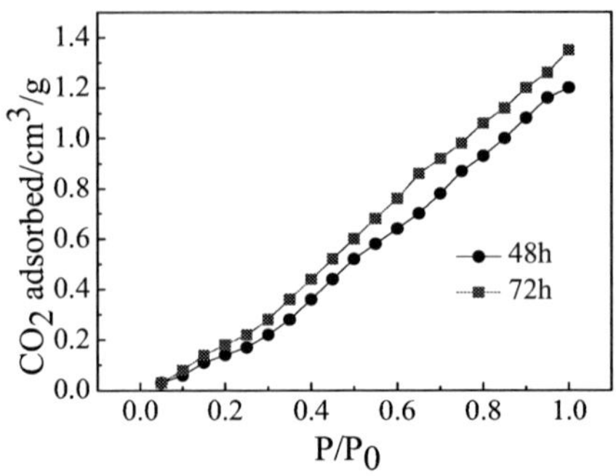

Fig.6 The carbon dioxide storage of sample synthesized for 48h and 72h

Conclusions

In conclusion, the zeolitic imidazolate frameworks synthesized by solvothermal synthesis method characterized with XRD showed that the crystallinity of the products synthesized by $Co(NO_3)_2{\cdot}6H_2O$ was lower than by $Zn(NO_3)_2{\cdot}6H_2O$. That increasing the molefraction of imidazole had a good effect on ZIFs with crystalline structure. The reaction time and temperature also had the great effect on the ZIFs syntheses. These ZIFs were in the form of single prism-shaped crystals by SEM. These porous materials have a capacity of 1.35ml/g for storing carbon dioxide at 273 Kelvin under ambient pressure.

References

1. H. Q. Yang et al., "Progress in Carbon Dioxide Separation and Capture: A Review," *Journal of Environmental Sciences-China*, 20 (2008), 14-27.
2. H. Y. Huang et al., "Amine-Grafted MCM-48 and Silica Xerogle as Superior Sorbents for Acidic Gas Removal from Natural Gas," *Ind. Eng. Chem. Res*, 42 (12) (2003), 2427-2433.
3. G. Ferey et, al., "Hybrid Organic-Inorganic Frameworks: Routes for Computational Design and Structure Prediction," *Angew. Chem*, 116 (2004), 6450-6456.
4. C. Jason et al., "Designing adsorbents for CO_2 capture from flue gas-hyperbranched aminosilicas capture of capturing CO_2 reversibly," *J. am. chem. soc*, 130 (2008), 2902-2903.

5. K. S. Park et al., "Exceptional Chemical and Thermal Stability of Zeolitic Imidazolate Frameworks," *Proc Natl Acad Sci. U.S.A*, 103 (2006), 10186-10191.
6. B. Wang, H. Furukawa, and O. M. Yaghi, "Colossal Cages in Zeolitic Imidazolate Frameworks as Selective Carbon Dioxide Reservoirs," *Nature*, 453 (2008), 207-212.
7. X. C. Huang et al., "Ligand-Directed Strategy for Zeolite-Type-Metal-Organic Frameworks: Zinc(II) Imidazolates with Unusual Zeolitic Topologies," *Angew. Chem. Int. Edn*, 45 (2006), 1557-1559.
8. Y. Q. Tian, Y. M. Zhao, and G. M. Zhang, "Design and Generation of Extended Zeolitic Metal-Organic Frameworks (ZMOFs): Synthesis and Crystal Structures of Zinc (II) Imidazolate Polymers with Zeolitic Topologies," *Chem Eur J*, 13 (2007), 4146-4154.
9. R. Banerjee, A. Phan, and O. M. Yaghi, "High-Throughput Synthesis of Zeolitic Imidazolate Frameworks and Application to CO_2 Capture," *Science*, 319 (2008), 939-943.
10. Takumi Higa, Masaki Moriya, and Yuichi Shimazaki, "Synthsis and Characterization of Imidazolate-Bridged Polynuclear Copper Complexes," *Inorganica Chimica Acta*, 360 (2007), 3304-3313.
11. C. S. Andrea, R. M. Andrew, and W. O. Nathan, "Design, Synthsis, Structure, and Gas Sorption Properties of Porous Metal-Organic Tetrahedral and Heterocuboidal Polyhedra," *J. am. chem. soc*, 127 (2005), 7110-7118.
12. Y. L. Liu et al., "Assembly of Metal-Organic Frameworks (MOFs) Based on Indium-Trimer Building Blocks: a Porous MOF with Soc Topology and High Hydrogen Storage," *Chem. Commun*, 46 (2007), 3278-3283.
13. Y. Q. Tian et al., "Determination of the Solvothermal Synthetic Mechanism of Metal Imidazolates via X-Ray Single Crystal Studies on the Photoluminescent Cadmium(II) Imidazolate and its Piperazine-Involved Intermediate," *Eur. J. Inorg. Chem*, 2004, 1039-1044.
14. Y. Q. Tian, C. X. Cai, and X. Z. You, "$[Co_5(im)_{10}\cdot 2MB]_\infty$: A Metal-Organic Open-Framework with Zeolite-Like Topology," *Angew. Chem. Int. Ed*, 41 (2002), 1384-1386.
15. Y. Q. Tian et al., "Two Polymorphs of Cobalt(II) Imidazolate Polymers Synthesized Solvothemally by Using One Organic Template DMA (N, N-Dimethylacetylamide)," *Inorg. Chem*, 43 (2004), 4631-4635.

Energy Technology 2010
Edited by: Neale R. Neelameggham, Ramana G. Reddy, Cynthia K. Belt, Ann M. Hagni and Subodh Das
TMS (The Minerals, Metals & Materials Society), 2010

Photochemical and Photo Electrochemical Conversion of Carbon Dioxide to Methanol Using Nanotubular TiO_2

M. Misra*, S. K. Mohapatra and K. S. Raja
Chemical and Materials Engineering/ M.S. 388
University of Nevada, Reno, NV 89557
* misra@unr.edu, 775-784-1603

Abstract

Ordered arrays of TiO_2 nanotubular photocatalysts were prepared by a simple anodization process. Band gap states of the TiO_2 were altered by controlling the anodization electrolyte composition and thermal annealing conditions that resulted in enhanced photo activity. These nanotubes were used for reducing CO_2 to methanol either by photochemical or photo electrochemical routes. Pd nano-particle loaded TiO_2 nanotubes were employed for photochemical reduction of CO_2 to methanol that showed a conversion rate of 12×10^{-3} mol $m^{-2}h^{-1}$. Carbon modified $TiO_{2-x}C_x$ type nanotubes were used as photo anode and Ti as cathode for photo electrochemical reduction of CO_2. The potential of Ti cathode was maintained at -0.5 V_{SCE}. The formation rate of methanol was determined as 13×10^{-3} $mol.m^{-2}h^{-1}$.

1. Introduction

Global warming due to emission of carbon dioxide is a serious environmental concern. The atmospheric concentration of CO_2 is about 384 ppm by volume and about 3×10^{12} tonnes by weight. Burning of fossil fuels and deforestation are the leading causes for increase in the anthropogenic CO_2. More than 32×10^9 tonnes of CO_2 are released per year (based on 2006 data) worldwide from the fossil fuel. Pumping CO_2 from the fossil fuel power plants into deep ocean basin is one of the most discussed disposal options. The other potential option is converting this green house gas into valuable fuels.

Carbon dioxide feed stock can be converted into fuels of solid, liquid or gas phase by chemical, electrolytic, photocatalytic and photo-electrocatalytic methods. Reduction of CO_2 to fuel form of any phase requires energy. The reduction process should be energy efficient so that we do not spend more energy and emit more (indirect) CO_2 in the process of converting CO_2. From this point of view, conversion processes occurring at room temperature look attractive.

Electrolytic conversion of CO_2 into various forms such as methane[1], ethane[2], ethylene[3], methanol[4,5], ethanol[6], n-propanol, formic acid[7], and formaldehyde has been widely reported. The standard equilibrium potential (at pH: 0) for reduction of CO_2 to methanol is 0.14 V_{SHE} and to methane is 0.2 V_{SHE} which is thermodynamically more favorable than hydrogen reduction. However, the gas phase reduction is kinetically complex requiring effective electrocatalyst. The major issues with electrolytic conversion of CO_2 to useful products are:

- Kinetic limitation owing to low solubility of CO_2 in water (0.033 mol/l at 25 °C, 1.06×10^5 Pa);
- Competition with hydrogen evolution reaction in aqueous solutions;

- Requirement of sp-metals such as Zn, Pb, Sn, In, and Cd or other oxide based electro catalysts as cathode in order to decrease the hydrogen evolution that leads to higher over potential and lower Faradaic efficiency; and
- A wide range of reaction products depending on the current density (non-selectivity)

The reduction of CO_2 to methanol requires six electrons following the reaction:

$$CO_2 + 5H_2O + 6e^- \rightarrow CH_3OH + 6OH^- \qquad (1)$$

The limiting current density for the above reaction will be about 200 mA/cm^2 for a concentration gradient of 0.33 mol/cm^2 following the expression:

$$i_L = nFDC_0/d_l \qquad (2)$$

Where, i_L = limiting current density, n = number of electrons (in this case 6), D = diffusion coefficient of CO_2 in electrolyte (assume 10^{-5} cm^2/s), C_0 = 0.033 mol/l, dl = diffusion layer thickness (10^{-1} cm).

If one could achieve such a limiting current density at a reasonably low over potential, electrolytic CO_2 reduction to methanol will be commercially viable. However, the required high over potential drives the electrolysis more toward hydrogen evolution. In order to make the process energy efficient, the energy required for the reduction process should preferably be supplied by renewable sources such as solar.

Photocatalytic reduction of CO_2 to formate using Pd-TiO_2 system has been reported[8]. Formation of formic acid at a rate of 0.28 μmol/h per g of TiO_2 catalyst has been reported with UV illumination in liquid CO_2 system.[9] Recently Wu et al[10] reported formation of ethylene by UV irradiation of TiO_2 particles loaded with 1 wt% Cu-Fe in CO_2 saturated water at 75 °C. However, the photo conversion efficiencies of the TiO_2 systems are very low (<0.1%). Kuwabata et al[11] reported a quantum efficiency of 5.9% for CO_2 to methanol conversion using ZnS micro crystallite with illumination of 280 nm light. The selectivity of the reduction product and higher conversion efficiency could be achieved by using appropriate electron donors such as methanol dehydrogenase, formate dehydrogenase, and pyrrollquinoline quinine as an electron mediator.

Photo electrocatalytic reduction of carbon dioxide has predominantly been reported in nonaqueous solutions, mostly dimethyl formamide (DMF). Bocris and co-workers[12] observed considerable reduction in over potential with CdTe photo electrode as compared to indium cathode in DMF with tetraalkyl ammonium salt as a supporting electrolyte and 18-crown-6 ether as catalyst. Recently Bocarsly[13] and coworkers reported selective under-potential conversion of CO_2 to methanol at an illuminated p-type GaP semiconductor in presence of pyridinium.

The overview of literature on the conversion of CO_2 to useful hydrocarbon indicates that a commercially viable electrochemical process for reduction of CO_2 to a selective fuel is still lacking. The quantum efficiency of TiO_2 thin film photo catalyst is very low. Wider band gap p-type semiconductors show promising results in the presence of electron mediators. An efficient photo electrochemical process requires development of a novel system consisting of a stable photo catalyst that can harvest the entire solar spectrum (if it is p-type, having active sites for CO_2 adsorption, high overpotential for H_2 reduction and low overpotential for CO_2 reduction, will be an added advantage), the conduction band minimum should be positioned above the LUMO of CO_2, electron donor/mediator, and an electrolyte system with high activity for CO_2. Currently no such system satisfying all these requirements is available. It is well known that nanoscale systems show enhanced photoactivity. For example, CdSe semiconductor surface in bulk scale do not participate in CO_2 fixation process. However, nanoscale CdSe below a critical size catalyze CO_2 fixation under illumination[14]. We have shown enhanced photo activity of ordered arrays of TiO_2 nanotubes as compared to TiO_2 thin films and nanoparticles[15,16,17]. The effect of TiO_2 nanostructures on the photo conversion of CO_2 to methanol or other fuels has not been explored in detail. This communication focuses on developing a commercially viable photoelectrochemical system based on functionalized TiO_2 nanotubular arrays for conversion of CO_2 to methanol using sunlight. Results of preliminary investigations of photochemical and photo electrochemical conversion of CO_2 to methanol are presented.

2. Experimental

2.1 Preparation of TiO_2 nanotubular photo catalyst functionalized with Pd nanoparticles:
Nanotubular TiO_2 arrays were formed by anodization of 0.2 mm thick Ti foils (size 70 x 50 mm^2) in fluoride (0.14 M NaF) containing 0.5 M phosphoric acid solution. A two-electrode configuration was used for anodization. Two larger Ti (75 x75 mm^2) sheets were kept on either side of the Ti sheet to be anodized and served as cathodes. The anodization was carried out at 20 V for about 1 h. After anodization, the samples were cleaned ultrasonically for about 45 seconds in ethanol solution, followed by rinsing in distilled water and dried in an electric oven at 100 °C for about an hour. Functionalization of TiO_2 nanotubes was carried out at this stage by loading with Pd. The dried TiO_2 samples were immersed in 0.5 wt% $PdCl_2$ containing ethanolic solution for 30 minutes under ultrasonication. The Pd containing solution wetted the internal and external surfaces of the TiO_2 nanotubes almost thoroughly because of pre-drying and ultrasonication. The Pd salt loaded samples were vacuum dried for about 12 h to remove ethanol. Then the samples were annealed at 500 °C for 2 h in a reducing atmosphere containing 10% hydrogen in argon. This heat treatment resulted in reduction of Pd(II) to Pd(0) nanoparticles loaded on to TiO_2 nanotubes. Furthermore, the amorphous TiO_2 nanotubes were transformed to crystalline (a mixture of anatase + rutile phases) nanotubes during the heat treatment at 500 °C. This photo catalyst is hereafter referred to as $PdTiO_2$ NT.

2.2 Preparation of carbon modified TiO_2 nanotubular arrays as photo anode:
Another set of experiments was carried out with TiO_2 nanotubes without any Pd loading. In this case, the titania nanotubes were prepared by anodization in ethylene glycol solution containing 0.14 M NH_4F and 2 vol% water at 20 V for 1 h under continuous ultrasonication. After anodization the samples were washed with ethanol, dried in an oven at 100 °C for 2 h and then annealed in 10% hydrogen + argon atmosphere at 500 °C for 2 h. Anodization of Ti in organic solvent followed by thermal annealing in reducing condition resulted in carbon modified TiO_2 nanotubes. These nanotubes will be referred to as $TiO_{2-x}C_x$.

The experiments on CO_2 reduction were carried out using two methods, viz 1) photocatalytic reduction using TiO_2 nanotubes functionalized with Pd nanoparticles; and 2) photo electrochemical method using $TiO_{2-x}C_x$ nanotubular samples as photo anode and Ti as photo cathode.

2.3 Photocatalytic reduction of CO_2:
Figure 1 shows the schematic experimental arrangement for reduction of CO_2 using TiO_2 nanotubular photocatalysts functionalized with nanostructured Pd. The functionalized photocatalyst in the form of a sheet was immersed in the CO_2 saturated dilute acid solution (pH 1.5) and was continuously illuminated with simulated solar light. The rate of conversion of CO_2 to methanol was determined by analyzing the solution at regular intervals using photo spectrometry. The pre-calibrated optical absorption data of known volume fractions of methanol in the solution were used for calculating the rate of methanol conversion in this investigation.

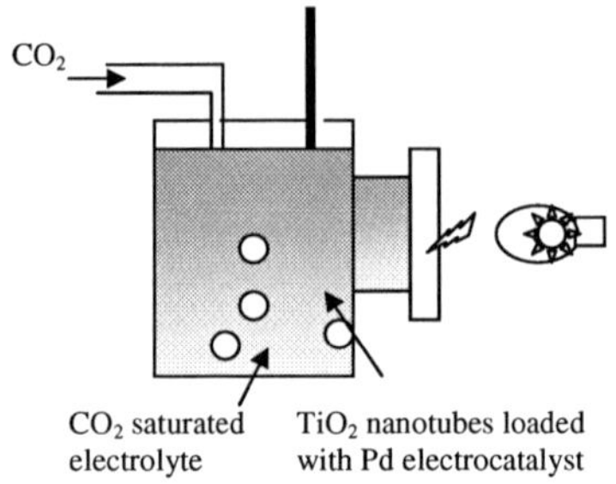

Fig. 1 Schematic experimental arrangement for photocatalytic reduction of CO_2.

2.4 Photo electrochemical reduction of CO_2: Photo electrochemical reduction of CO_2 was carried out using a three-electrode configuration in dilute sulfuric acid solution with pH 1.5. The electrochemical cell had a 3 inch diameter quartz window for light illumination, as shown in Figure 2. The PTFE lids of the cell had provisions for inserting working, counter and reference electrodes, gas inlet and gas out let. Commercial grade CO_2 was continuously purged in the solution at least one hour before start of the experiments to make sure CO_2 saturation. The gas purging continued in the cathode compartment during the experiments. Cyclic voltammetry and potentiostatic studies were carried out to evaluate the kinetics of CO_2 conversion using TiO_2 nanotubes as photo anodes. The cathode was a Ti foil. The cathode was placed in a separate compartment and was connected to the anode compartment through a porous frit (Ace Glass type E pores). The TiO_2 nanotubular samples were illuminated with a 300 W solar simulator (Model: 69911, Newport-Oriel Instruments, Stratford, CT, USA). A computer-controlled potentiostat (Model: SI 1286, Schlumberger, Farnborough, England) was employed to control the potential and record the photocurrent generated. The intensity of the light was measured by a thermopile sensor (Model 70268, Newport) using a radiant power and energy

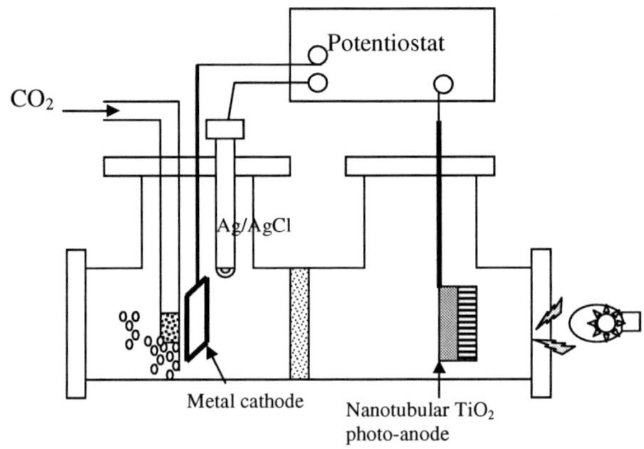

Figure 2 Schematic illustration of the electrochemical cell assembly used for reduction of CO_2.

meter (Model 70260, Newport Corporation, Stratford, CT, USA). Potential control was employed both for photo anode and Mo cathode. Electrolyte in the cathode compartment (where CO_2 was continuously purged) was extracted at regular intervals and analyzed for the presence of methanol. UV-Vis photo spectroscopic analysis was carried out to detect the presence of methanol in the electrolyte. 1 ml of solution was analyzed at every 30 minutes. In order to quantify the methanol volume fraction of the electrolyte, several calibration runs were carried out using known volume fractions of methanol in the electrolyte. The intensity of unfiltered light in all the experiments was ~ 246 mW/cm^2.

3. Results and Discussion

Figure 3(a) shows the TEM image of a TiO_2 nanotube decorated with Pd nanoparticles. The nanotubes prepared by anodization in acidified fluoride solution has inside diameter in the range of 80 – 100 nm and a length of about 400 - 600 nm. The wall thickness of the nanotubes was in the range of 15 – 25 nm. The Pd nanoparticles were found to be present uniformly on the walls of the nanotubes with a size distribution ranging from 3 -14 nm. The weight percent of the Pd nanoparticles loaded onto the TiO_2 nanotubes was determined by TEM-EDX analyses of several nanotubes. The loading varied in the range of 0.8 - 1.2 wt% between the nanotubes present in a coupon. Our previous study indicated that 1.25% Pd loading was the optimum condition for complete photo degradation of methyl red azo dye[18]. Therefore, a similar Pd loading concentration was used in the present investigation also.

Figure 3 (b) shows the top view of the nanotubular TiO_2 array prepared by anodization in ethylene glycol based solution under ultrasonicated condition. The diameter of the nanotubes prepared in organic solvent based electrolyte was smaller than (60 – 80 nm) that prepared in fluorinated phosphoric acid. However, the wall thickness was not significantly different. The length of the nanotubes was a strong function of anodization time in case of the organic based electrolytes. Figure 3 (c) shows the side view of the TiO_2 nanotubular array. The length of the nanotubes employed in this investigation was in the range of 800 – 1000 nm. It should be noted that several micrometers long (> 60 μm) nanotubes can be grown and used as photo anodes. Our experience indicated that when the nanotubes were more than 60 μm long, the Ohmic contact between the TiO_2 layer and the Ti substrate was poor after thermal annealing. Several reasons could be attributed to this effect. A predominant factor that affects the Ohmic contact could be the rate of heating and cooling of the thermal annealing. Furthermore, factors such as growth stresses during anodization, annealing environments, purity of the ethylene glycol, purity of the Ti foil and dimensions of the anodized coupons affect the stability of the Ohmic contact. Anodization of both sides of the coupon (as against only one side by masking the other side with tape or epoxy) was observed to minimize exfoliation of the TiO_2 nanotubular layer.

Figure 4 shows the UV-Vis photo spectroscopic results of the CO_2 conversion process using photocatalytic and photo electrochemical conversion processes. The absorption spectrum of the CO_2 saturated solution is almost flat (labeled as blank). Addition of 0.75 mM methanol in the blank solution gives absorbance peak at about 190 nm.

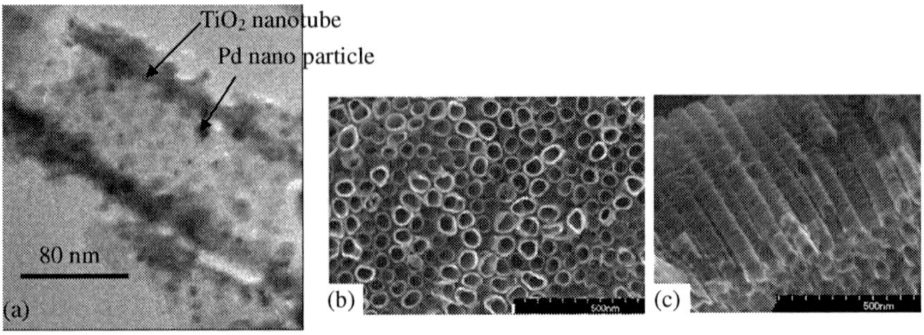

Figure 3 (a) TEM image of a TiO_2 nanotube loaded with Pd nanoparticles; (b) FESEM image of the top surface of the $TiO_{2-x}C_x$ nanotubes; and (c) side view of the array of $TiO_{2-x}C_x$ nanotubes.

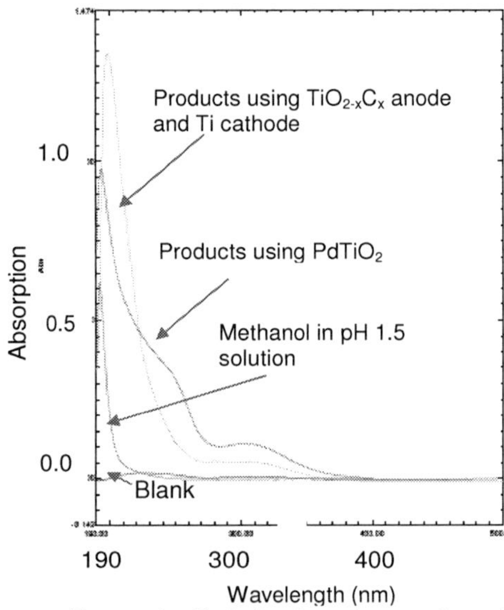

Figure 4. Optical absorption results of determination of methanol in the electrolyte by converting CO_2 using photocatalytic and photo electrochemical methods

Table.1 Methanol production from CO_2 using semiconductor catalysts and solar light.[a]

Catalysts	Conditions	Methanol yield[b]
No catalyst	CO_2, solar light	No product
$TiO_{2-x}C_x$ NT	CO_2, $TiO_{2-x}C_x$ anode, Ti cathode, -0.5V_{SCE}, solar light	1.6 x10^{-3} mol/hr
$PdTiO_2$ NT	CO_2, $PdTiO_2$, solar light	1.2 x10^{-3} mol/hr

[a]Experiments are carried out in aqueous H_2SO_4 solution (pH 1.5).
[b]Yield is calculated from UV spectroscopy.

Table 1 summarizes the optical spectroscopy results obtained in different experimental conditions. The rate of methanol formation using $PdTiO_2$ NT catalyst was calculated to be 1.2 mM/hour. The geometrical area of the photocatalyst coupon was 35 cm^2. It should be noted that because of the nanotubular surface morphology, the actual surface area of the TiO_2 would be much larger. Based on the dimensions of the nanotubes, the actual surface area was increased by a factor of 23.4. Therefore, the estimated surface area of the TiO_2 was about 820 cm^2. The TEM images of the $PdTiO_2$ NTs suggested a surface coverage of > 35% by the Pd nanoparticles. Since the Pd nanoparticles had a wide range of dimensions and irregular shapes, calculation the real surface area was difficult. Nevertheless, the total area of the $PdTiO_2$ NT photocatalysts of this investigation would be around 900 – 1000 cm^2. Therefore, the methanol formation rate using $PdTiO_2$ NT photocatalyst could be estimated as 12 x10^{-3} mol $m^{-2}h^{-1}$.

Figure 5 schematically explains the anodic and cathodic reactions occurring at different sites of $PdTiO_2$ NT. When the TiO_2 nanotubular array is illuminated with solar light, electron-hole pairs are generated. Holes are consumed by the water oxidation reaction (3).

$$3H_2O + 6h^+ \rightarrow 6\,H^+ + 1.5\,O_2\uparrow \quad (3)$$

During this process H^+ ions are generated. The electrons are trapped by the Pd nanoparticles and participate in the CO_2 reduction process (4)

$$CO_2 + 6\,H^+ + 6e^- \rightarrow CH_3OH + H_2O \quad (4)$$

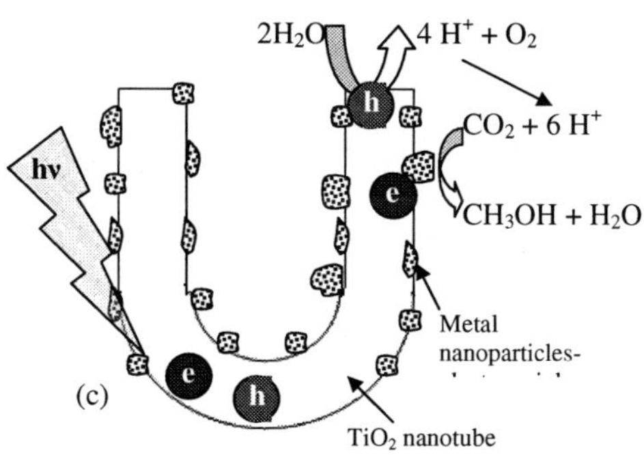

Figure 5. Schematic illustration of reaction sites for reduction of CO_2 to methanol. Oxygen evolution preferentially occurs at TiO_2 as anodic reaction and the corresponding cathodic reaction of methanol formation is considered to occur at Pd sites.

It should be noted that the other cathodic reaction could be hydrogen evolution. Since Pd has very high exchange current density for hydrogen evolution, this reaction could occur more preferentially than the CO_2 reduction. However, in this present investigation, formation of a significant quantity of methanol indicated that the hydrogen evolution reaction did not detrimentally affect the reduction of CO_2. It is possible that the CO_2 reduction occurred in multiple steps and presence of H_2 assisted the conversion of intermediate products to methanol. Olah et al[19] proposed that TiO_2 could act as catalyst for dimerization of formaldehyde to give methyl formate. Hydrogenation of methyl formate gives methanol as shown in the following reactions:

$CO_2 + 4H^+ + 4e^- \rightarrow HCHO + H_2O$ (5a)

$2CH_2O \rightarrow HCOOCH_3$ (5b)

$HCOOCH_3 + 2H_2 \rightarrow 2CH_3OH$ (6)

The redox potential for the reaction (4) is around -0.46 V_{SCE} at pH 1.5. During photo electrochemical reduction experiments, potential of either TiO_2 nanotubes (photo anode) or Ti cathode was controlled at a desired value. In case of anodic potential control, the potential of the $TiO_{2-x}C_x$ NT was maintained at 0.9 V_{SCE} which was the redox potential for oxygen evolution in the pH 1.5 electrolyte. In case of the cathodic potential control, the potential of the Ti cathode was maintained at -0.5 V SCE. Almost similar results were observed in either of the potential controls under light illumination of the $TiO_{2-x}C_x$ NT. The photo potential developed by $TiO_{2-x}C_x$ NT was about 0.65 V. The photo current density was observed to be in the range of 2.3 – 2.8 mA/cm^2. The formation rate of methanol was observed to be 1.6 x 10^{-3} mol/h for the electrode area of 35 cm^2. Considering the dimensions of the nanotubes, the estimated apparent surface area was about 35 times that of the geometric area. Therefore, the normalized methanol formation rate could be estimated as 13 x 10^{-3} mol $m^{-2}h^{-1}$ which is slightly higher than the rate observed by photochemical route using $PdTiO_2$ NT.

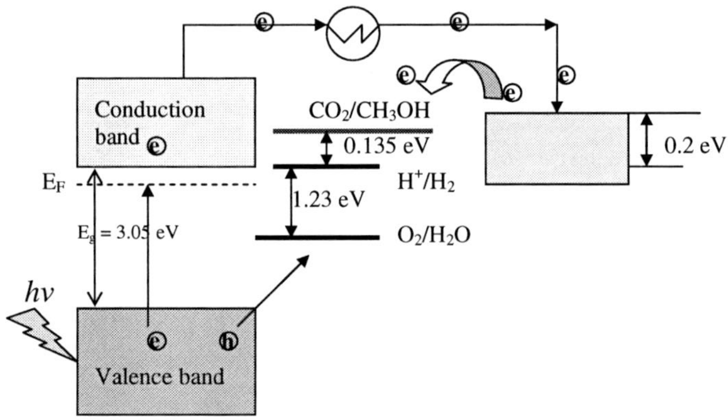

Figure 6. Energy level of CO_2 reduction to methanol in a low pH electrolyte in comparison with energy level of hydrogen evolution using TiO_2 photo anode. The cathode is externally polarized to transfer the electron for CO_2 reduction.

Figure 6 illustrates the energy levels of the oxygen and hydrogen evolution, and CO_2 reduction with reference to the energy gap of the TiO_2 n-type semiconductor. Since the conduction band minimum of the TiO_2 is much lower than the energy level of CO_2/CH_3OH, an external bias is required for electron transfer to complete the CO_2 reduction reaction. From the energy diagram, it is clear that hydrogen reduction is the energetically favorable reaction. However, hydrogen evolution can be kinetically limited by choosing a cathode that shows a high over-potential for H_2 reduction. In this case, Ti was used as cathode since the exchange current density for H_2 reduction onto Ti was about 5 orders of magnitude lower than that of Pd. In this investigation, faster kinetics of CO_2 reduction to methanol was observed without applying significant over potentials. The required over potential was supplied by the photo catalytic effect. It should be noted that the methanol formation rate calculations were based on the UV-Vis photo spectroscopic results. Further studies are being carried out to analyze the exact concentration of methanol using mass spectroscopic techniques. When compared to the photochemical method, the photo electrochemical method showed a slight increase in the conversion rate. However, significant electrical energy was supplied in this process. On the other hand, expensive Pd nanoparticles were used as catalysts in the photochemical method. Long term experiments and more deterministic analyses are required to compare the stability and efficiency of both these processes.

4. Conclusions

Ordered arrays of TiO_2 nanotubes loaded with and without Pd nanoparticles were used as photo catalyst and photo electrocatalyst for reduction of CO_2 to methanol. Preliminary investigations indicated that both these materials show promising results. The methanol formation rates were determined by comparing the UV-Vis spectroscopic results with a standard calibration. Pd nano-particle loaded TiO_2 nanotubes were employed for photochemical reduction of CO_2 to methanol that showed a conversion rate of 12×10^{-3} mol $m^{-2}h^{-1}$. Carbon modified $TiO_{2-x}C_x$ type nanotubes were used as photo anode and Ti as cathode for photo electrochemical reduction of CO_2. The potential of Ti cathode was maintained at -0.5 V_{SCE}. The formation rate of methanol was determined as 13×10^{-3} mol.$m^{-2}h^{-1}$. Further studies are required to understand the stability of the catalysts and sustainability of the conversion reactions.

5. References

[1] R. L. Cook, R. C. MacDuff, A. F. Sammells, J. Electrochem. Soc., 140 (1993) 3069 - 3071

[2] Y. Hori, K. Kikuchi, A. Murata, S. Suzuki, Chem. Lett., (1985) 1695

[3] K. Hara, A. Kudo, T. Sakata, M. Watanabe, J. Electrochem. Soc., 142 (1995) L57-L59

[4] A. Bandi, J. Electrochem. Soc., 137 (1990) 2157-2161

[5] M. Schwartz, R. L. Cook, V. M. Kehoe, R. C. MacDuff, J. Patel, A. F. Sammells, J. Electrochem. Soc., 140 (1993) 614 - 618

[6] A. Bandi, H.-M. Kuhne, J. Electrochem. Soc., 139 (1992) 1605-1610

[7] P.G. Russell, N. Kovac, S. Srinivasan, M. Steinberg, J. Electrochem. Soc., 124 (1977) 1329-1338

[8] V. Heleg, I. Willner, J. Chem. Soc., Chem. Commun.(1994) 2113

[9] S. Kaneco, H. Kurimoto, K. Ohta, T. Mizuno, A. Saji, J. Photochem. Photobiol, A: Chem 109 (1997) 59

[10] T. V. Nguyen, J. C. S. Wu, App. Catalysis A: General 335 (2008) 112-120

[11] S. Kuwabata, K. Nishida, R. Tsuda, H. Inoue, H. Yoneyama, J. electrochem. Soc., 141 91994) 1498-1503

[12] J. O'M. Bocris, J. C. Wass, J. Electrochem. Soc., 136 (1989) 2521-2528.

[13] E.E. Barton, D. M. Rampulla, A. B. Bocarsly, J. Am. Chem. Soc., 130 (2008) 6342 - 6344

[14] L. G. Wang, S. J. Pennycook, S. T. Pantelides, Phys. Rev. Lett., 89 (2002) 075506-1

[15] K. S. Raja, M. Misra, V. K. Mahajan, T. Gandhi, P. Pillai, S. K. Mohapatra, J Power Sources, 161, (2006) 1450-1457

[16] S.K. Mohapatra, M. Misra, V.K. Mahajan, K.S. Raja, J. Catal. 246 (2007) 362-369

[17] S. K. Mohapatra, Mano Misra, J. Phys. Chem. C, 111, , (2007) 11506-11510

[18] S. K. Mohapatra, N. Kondamudi, S. Banerjee, M. Misra, Langmuir, 24 (2008) 11276 - 11281

[19] G. A. Olah, A. Goeppert, G. K. S. Prakash, J. Org. Chem., 74 (2009) 487-498

Energy Technology 2010:
Conservation, Greenhouse Gas Reduction and Management, Alternative Energy Sources

Energy Conservation in Metals

Organizers

Cynthia Belt
Mark Cooksey
Donald Whipple
Russell Hewertson

Energy Technology 2010
Edited by: Neale R. Neelameggham, Ramana G. Reddy, Cynthia K. Belt, Ann M. Hagni and Subodh Das
TMS (The Minerals, Metals & Materials Society), 2010

FIVE LOW COST METHODS TO IMPROVE ENERGY EFFICIENCY ON REVERBERATORY FURNACES

Cynthia K. Belt[1], Ray D. Peterson[2], and David Bequette[1]

[1]Superior Industries International, 1901 Borick Drive, Fayetteville, AR 72701, USA
[2]Aleris International, 397 Black Hollow Rd, Rockwood, TN 37854, USA

Keywords: Energy, Furnace

Abstract

With a need to cut production costs in the metals industry, cost savings from energy efficiency improvements are critical. Opportunities can be found by reviewing furnace systems and practices so that the melting and holding processes operate at the highest energy efficiency. Many of the changes can be implemented with a minimum of expense or capital investment. This paper will review several low cost methods to optimize energy efficiency on reverberatory melting or holding furnaces. These five methods (furnace utilization, idle procedures, fire rate, controls, furnace pressure) will be explained and information given to implement these changes. Included are several rules-of-thumb and benchmarks for aluminum reverb furnaces.

Introduction

Energy can be a significant portion of the total manufacturing costs for metals and materials manufacturing involving thermal processing such as heat treating, melting, and high temperature processing. Natural gas prices have dramatically varied over the last few years. While no one can foresee the future, prices are expected to increase from the current low prices. More recently, emphasis is being put on the reduction of greenhouse gases with the potential of additional costs to industry.

One of the first steps in an energy management program is to measure the energy usage of the process [1]. The results are typically understood in relation to the mass melted in kWh / mT [BTU / lb]. For the purposes of this paper, furnace energy efficiency is calculated over a month or even a year and thus includes hold times. This efficiency is typically different from the values quoted by furnace manufacturers. Those values are for limited short term tests at high utilization.

Once you know the performance for your furnaces, these should be compared to industry benchmarks. For aluminum melt furnaces with cold air burner systems, a benchmark value or goal is 1800-2000 BTU/lb [1160-1290 kWh/mt], though some of the best have reached 1600 BTU/lb [1030 kWh/mt]. Furnaces with regenerative burners can attain 1100-1200 BTU/lb [710-775 kWh/mt]. There are some designs of multi-chamber furnaces or oxygen furnaces that have improved on these numbers, perhaps as low as 950 BTU/lb [615 kWh/mt]. Typically, as the degree of sophistication and energy efficiency of the furnace increases, so does the cost.

While these high capital technologies like regenerative burners or multi-chamber furnaces can make major improvements in energy efficiency, these are often difficult to cost justify with current low energy prices or for lower throughput. For instance, foundries within the casting industry tend to utilize smaller furnaces with relatively low throughput. For both these low energy users and as a first step for high energy users, a review of low cost energy efficiency improvements should be explored.

Topics discussed within this paper include (1) furnace utilization, (2) idle procedures, (3) fire rate, (4) controls, and (5) furnace pressure. Maintaining the proper air / fuel ratio, low temperature combustion air preheat, and improved insulation are other low cost methods to optimize energy efficiency. These are not included in this discussion, but can be found in several other resources [2,3,4,5].

The majority of the benchmarks and rules-of-thumb within this paper apply to aluminum sidewell melt furnaces. As with all rules-of-thumb, they are strictly a guide. The optimum numbers are affected by many items: furnace design, burner placement, burner velocity, stirring method and furnace turnover, furnace practices, etc. The best method for optimizing a current furnace is by either running a test or using computer modeling to investigate multiple design changes. However, using these rules-of-thumb can help in identifying potential changes. The authors have used these rough calculations to evaluate changes in over 150 furnaces.

Furnace Utilization

Even the best furnace will perform poorly, if underutilized. As mentioned in earlier papers [6,7] and as seen in Figure 1, energy efficiency greatly drops off below 50 % utilization. Even at the 50 % point, the furnace is 33% less energy efficient than 100 % utilization. To remain within 10 % of the furnace capability, it must run at 80% utilization on average for the month. Overall, this 80% utilization value for the month is a good goal since the furnace will require time for maintenance, skimming, and other activities.

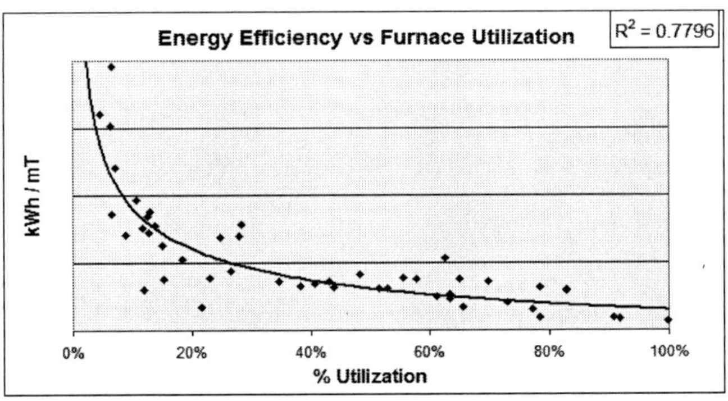

Figure 1. Energy Efficiency vs. % Utilization

First, however, we must understand what is the maximum throughput for a given furnace. This may not be the value given by the manufacturer. In some cases the manufacturer's stated melt rate is 'low-balled' to insure the furnace will meet the customer requirements. One method is to track the daily throughput and energy efficiency over an extended time and plot it. Does the data show a leveling off of energy efficiency similar to the graph above? If not, the furnace may not be fully utilized. If is does level off, the highest daily value could be used as 90% utilization.

Calculations based on the furnace design may also be used. In a wet bath reverberatory furnace, heat is transferred into the metal through the bath surface within the main chamber. This heat transfer tends to be fixed so that a multiplier can be used. The heat transfer value is dependent on the furnace technology. Table I gives values for three different furnace designs. Multiplying this value by the bath area will give a good approximation of the capability of the furnace.

Table I. Evolution of Melter Reverb Furnace Design [8]

Property	Uncirculated	199X Design	2000X Design
Heat Transfer	38 lb / ft^2	56 lb / ft^2	70 lb / ft^2
Burner Type	Cold Air	Cold Air	Regenerative
Pump Turnover	None	4X per hour	8X per hour
Feed	Cold	Cold	2/3 Preheated, 1/3 Cold
Charge Method	Choke	Batch	Continuous
Energy Efficiency	3000 BTU/lb	2000 BTU/lb	1000 BTU/lb

The next question is why the furnace is not fully utilized. Low production needs may require shutting down extra furnaces. Often, the plant management is used to having several furnaces running when production needs are high, but then forget to analyze the costs for the extra furnace when production demands change. Another method is to fully utilize the furnace during certain shifts and implement idle procedures when production is not needed (see next section).

Utilization may also be affected by the practices surrounding the furnace. These include charging, skimming, maintenance, and casting practices. Some points here:

- The optimum charging method is a continuous charge using some sort of mechanical means. If that is not reasonable, charging on a 15 minute basis will improve throughput.
- Skimming practices should be examined. In general, once per 8 hour shift is satisfactory for a side bay melter. If more skimming is needed, there may be other problems with the furnace that is increasing dross formation [9].
- A time limit should be set so that skimming is complete, but door open time is reduced.
- Charging and skimming should be done quickly to minimize door open time.
- The door should not be opened unless everything is staged and ready.
- Door opening systems in the cab for the loader operator can decrease door open time.
- The maintenance of the door system should not be overlooked. If the door is slow to open, heat loss will increase. In general, the door should fully open in 20 seconds or less.
- Finally, casting should be as even as possible. The goal is the make the rate of metal into the furnace equal to the rate of metal out of the furnace. Any bottlenecks in the next process will affect the furnace and should be reduced.

Idle Procedures

Most melt furnaces will need to spend some time idling. This can occur due to production needs, manning requirements, or for maintenance. Settings of the furnace should be altered to optimize the furnace for this holding period. Every thermal process is different in setting this procedure. It must be done at the plant level knowing the equipment and the process. These settings include:

1. Bath temperature - Depending on the alloy, bath temperature setpoint can be set down to as low as 1225° F [663° C] with metal still in the furnace, though higher temperatures will still save energy (see control section for savings). Time must be allowed to bring the bath up to temperature for production (~1-2 hrs.).
2. Roof or flue temperature - For furnaces with cascade logic, reducing the flue or roof temperature will reduce heat loss through the flue. A setpoint of 1400° - 1600° [760 - 870° C] is recommended for reverb furnaces.
3. Fire rate - When the controls allow, the fire rate should be reduced to just maintain holding temperatures. In some furnaces, shutting off certain burners can help. This has been suggested to keep furnace velocity and heat transfer high.[13]
4. Well covers - Any furnace opening should be closed when possible. This may involve using fiber blankets or building a cover to set on an open well area.
5. Furnace pressure – Furnaces are positively pressured to avoid drafting that will pull cold air into the furnace. If the fire rate was decreased during idling, maintaining furnace pressure is particularly important. If there is no damper, the flue area can be reduced in size with steel or refractory board. See the furnace pressure section for calculations.
6. Pump Setting – Reducing pump speed should reduce dross production. Some stirring is still preferred.
7. External blowers and fans - When possible, external blowers should be shut off to baghouses. This reduces electricity use and reduces heat loss from the furnace. Follow proper procedures as required by environmental regulations. During hot weather, make sure that personnel fans are not blowing toward the furnace walls.

Ideally, software programming can be written to make these changes with a single press of the button. However, these items can be quickly completed within 10-15 minutes and are more than paid back in savings. On a daily idled basis, savings have been up to 67%. During these furnace idle times, the furnace is running as a holding furnace. White gives a value of 25 BTU/lb/hr [10] [16 kWh/mT/hr] for holding furnaces. As an example, for a furnace with 50,000 pounds this would be 30 MM BTU per day. This value can be used as a goal for gas usage on down days.

Controls

Control optimization can include settings, practices, or the physical control systems. The controls can include temperature control and air / fuel ratio control.

Critical to energy efficiency is to make sure additional energy is not applied when it is not needed by the process. While this step sounds obvious, it is often overlooked or suffers from the "we've always done it this way" syndrome and requires a fresh outlook. Depending on the current controls of the furnace, this change can be very simple or may require some low cost capital.

First, understand the current settings of the furnace, the range of the metal temperature required for casting, and the requirements of subsequent processes. If there are no measurements available, take several readings at different times and stages to understand the current state.

1. Is there any heat loss occurring that can be reduced? Examples of this might include better insulation, adding covers, a shorter trough run, or reducing delivery and holding times.
2. Is there any hidden heat loss? Make sure you are getting the real story. The authors have

seen instances where operators were opening the doors of the furnace to lower the temperature to be able to cast.

3. Does the next process need the temperature as delivered? Sometimes additional cooling is added at the next process when the better solution is to never add the heat at all! Casting hot may result in longer periods for ingots to solidify or more energy might be used from a cooling tower to cool molds that are too hot.

The potential of savings from lowering the bath setpoint can be estimated by:

$$\text{Savings} = \Delta T \times c_p / \text{eff} \qquad [1]$$

Where Savings = savings in energy intensity (kWh / T) [BTU / lb]
ΔT = difference in temperature (C) [F]
c_p = specific heat of molten metal (kWh/T/° C) [BTU/lb/° F]
eff = thermal efficiency of the furnace

For example, for a furnace with aluminum where the average bath temperature is changed from 1400° F to 1390° F and the furnace normally runs at 25% efficiency (~2000 BTU/lb), the savings is:

$$14.7 \text{ BTU / lb} = 10° \text{ F} \times 0.368 \text{ BTU/lb/° F} / 25\% \qquad [2]$$

There may be additional benefits with shorter melt times, reduced oxidation, and improved metal quality from reduced hydrogen pickup.

On some systems, capital may be needed to improve the controls. The minimum for any furnace is to have the burners controlled by the bath temperature. Cascade controls are a good next step. The terminology "cascade controls" has been used for various control schemes. Here it means to use both the bath temperature and either the roof or flue temperature to control the burners. The burners are controlled by the bath temperature until the point where the roof temperature reaches its set point. At this point the burners will back down. This reduces the flywheel affect that after the bath temperature is attained; the heat within the walls can cause the bath to overshoot.

Data collection and graphing will provide benefits. The well-used phrase "If you can not measure it, you can not improve it." applies here. By tracking data, the process can more easily be improved. Data collection allows for troubleshooting problems over time. Suggestions for data collection include: bath temperature, roof temperature, flue temperature, fire rate, door opening time, pump speed, transfer time, and furnace pressure.

It is beneficial to graph the bath temperature over time. Figure 2 shows an example where the bath temperature overshot the set point 12.9% of the time within a week. Sometimes graphing this by shift will show differences between operators. Overshoot can be reduced by several methods including a lower roof set point, PID (proportional-integral-derivative) settings, or constant charging/casting.

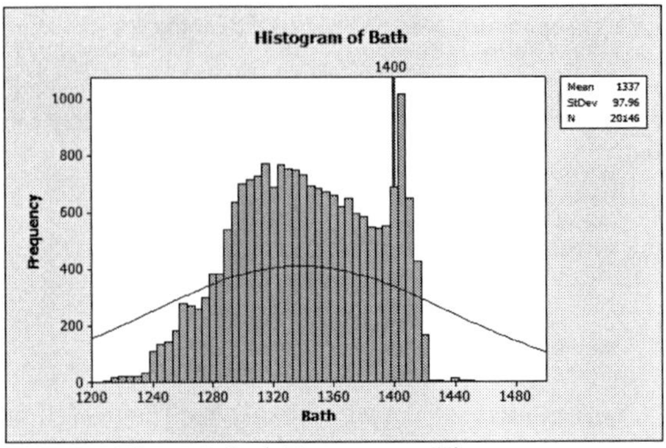

Figure 2. Bath Temperature Histogram

The burner air / fuel controls should also be considered. While many furnaces successfully use low cost cross-connected systems, mass-flow is a better control scheme and is cost justified in many instances. Mass-flow directly measures the air and gas flows, then adjusts them to meet the required air / fuel ratio and fire rate. Since it monitors the temperature of the gases, these controls in particular will help with seasonal changes, systems where air temperature may vary, and is highly suggested for combustion preheat. A rough estimate is that there will be a 5 - 10 % improvement by adding this system.

Fire Rate

Many furnaces have been over-designed in regards to burner fire rate. This extra safety factor may ensure the highest melt rate, but in reality the furnace will be less energy efficient while the extra fuel may not be increasing the melt rate. The molten bath surface can only absorb heat at a certain rate. Exceeding the rate where heat is absorbed by the molten bath surface, extra heat is lost up the flue. In extreme cases, a flame can be seen in the flue when this occurs.

When the fire rate is too high, the furnace will run at higher BTU/lb and if the fire rate is too low, melt rate can be affected. A median value must be found. The first step in optimizing the fire rate is to determine the square footage of the bath within the main chamber. The square footage is multiplied by a certain value to calculate the optimum high fire rate of the total burner input. In the past, 100,000 BTU/hr/ft^2 was used for firing density [11]. However, the same paper recommended lowering this to 80,000 BTU/hr/ft^2 for high velocity burners. Based on computer modeling and furnace testing, a value of 66,250 BTU/hr/ft^2 is now recommended [4]. It is best to test at different fire rates to optimize both energy efficiency and match the desired melt rate. Data suggests that this value may depend on the flame temperature and velocity and a host of other factors.

Holding furnaces appear to be optimized at around 43,000 BTU/hr/ft^2, though as low as 20,000 BTU/hr/ft^2 [11] and 10,000 BTU/hr/ft^2 [13] has been recommended. This recommendation of an "ideal" fire rate does not appear to work for round top furnaces or dry hearth furnaces. On these furnaces, heat is transferred directly to the solid aluminum and not into the molten bath. The shape and density of the charged aluminum will affect the heat transfer for these types of furnaces.

In a study of approximately one hundred reverb furnaces with multiple manufacturers, 80% of the melting furnaces were designed to run over the recommendation of 66,250 BTU/hr/ft^2 Even within holding furnaces, 31% were oversized on their fire rate using the melt furnace value of 66,260 BTU/hr/ft^2. By reducing the fire rate, savings have varied between 5-15 %.

Furnace Pressure

While the evils of negative pressure in a furnace are well known, the condition still exists in many furnaces within the industry. Furnace designers purposely oversize the flue to ensure a safety factor and prevent over pressurization [12]. Yet, negative pressure will pull in cold air from the environment. The cold gas can cool metal that reduces energy efficiency, cool the burner flame causing incomplete combustion, and add oxygen to the chamber which can increase dross production.

Furnace pressure is typically measured at 1 - 2 feet [0.3 - 0.6 m] above bath level. A pressure port can be added at this level to monitor the pressure with reduced risk of metal splash filling the hole. A common value within industry is to maintain +0.02 inches water column [0.05 mb] at this point [12].

To size the flue correctly, there are tables and charts [14] or software programs available [15]. Flue size is dependent on stack height, flue gas temperature, and pressure drops due to horizontal runs, flue shape, materials, etc. However, to roughly size the flue, a rule of thumb is that a typical aluminum reverb flue can support 20,000 BTU/hr/in^2 or the inverse of 50 in^2 of flue area for each million BTU of burner input [16]. The most accurate method to understand furnace pressure for an existing furnace is to just check it!

A quick method to check furnace pressure is to hold a narrow strip of paper several inches from the door at a point where there is a slight opening due to refractory or door damage. At high fire, the strip should be pushed out. Negative pressure will pull the paper in toward the furnace. Warning: if the furnace pressure is a too high this can be dangerous due to the heat and possible flames. A more accurate method is to use a hand-held manometer with a tube inserted into the closed furnace approximately one foot.

At a minimum, the furnace should be positive at high fire. If not, the flue is oversized. This can be easily remedied by covering the furnace with plates made of carbon steel, stainless steel, or refractory boards. The next step is to install a partial damper for idle times when burners are shut off. This is a low cost plate that can be moved into and out of place quickly. Warning: the plate must only be used when the furnace is prevented from going to full fire and must be removed before returning to normal operation. Finally, the best solution is to install a flue damper and with controls that monitor the furnace pressure. These systems are fairly low capital and can be cost justified in many furnaces within 1-3 years. Savings for good furnace pressure control has been given at 7% [2], though a range of 5-10 % can be used.

Conclusions

Cost savings are possible within aluminum furnaces by following the suggestions within this paper on furnace utilization, idle procedures, fire rate, controls, and furnace pressure. While these points are not new, a new generation of engineers must be trained in the basics. Both older and newer furnaces can benefit from a review of the basics for energy efficiency. The rules-of-thumb are not meant to be utilized by furnace designers, but by engineers and others within a foundry to evaluate when further testing may optimize a system. These items can be implemented with little to no capital, provide substantial savings, and should be a first step before the more expensive improvements are considered.

References

1. Ray D. Peterson and Cynthia K. Belt, "Elements of an Energy Management Program", Vol. 61 No. 4, JOM, April 2009, pg 19-24

2. "Energy Tips, Tip Sheet #2, Check Burner Air to Fuel Ratios", www1.eere.energy.gov/industry/best practices, Office of Industrial Technologies, Energy Efficiency and Renewable Energy, U.S. Department of Energy

3. North American Combustion Book, Volume I, 3rd Edition, 1997

4. Cynthia K. Belt, Brian M. Golchert, Paul E. King, Ray D. Peterson, Joseph L. Tessandori, "Industrial Application of DOE Energy Savings Technologies to Aluminum Melting", Light Metals 2006, Travis J. Galloway Ed., TMS, 2006, pp. 881-885

5. David White, "Energy Savings Myths", Die Casting Engineer, January 2009, pg 36-39

6. Cynthia K. Belt and Ray D. Peterson, "Analysis of Energy Efficiency for Industrial Processes", EPD Congress 2008, S. Howard Ed., TMS, 2008, pp. 53-60.

7. C. Edward Eckert, Mark Osborne, Ray D. Peterson, "Improving Energy Efficiency in a Modern Aluminum Casting Operation", Energy Technology Perspectives, Neelameggham et al Ed., 2009

8. Conversation with Rich Henderson, Metallics / Pyrotek.

9. David White, "Fuel Saving Techniques", H.E.L.P. Newsletter, Vol. 2 Issue 3, March 2009

10. David White, "Operating Furnaces Efficiently", webinar for American Foundry Society, 2009

11. John A. Marino, "High Efficiency Aluminum Melting, European Experience and Background", presented at the IHEA Energy Seminar, 1982

12. Eclipse Combustion Engineering Guide, 1986, EFE 825, 8/04, pg. 53

13. John S. West, "U.S. Experiences with High Efficiency Aluminum Melting Techniques", presented at the IHEA-AGA Energy Seminar, 1983

14. North American Combustion Handbook Volume II, 3rd Edition, 1997

15. Hauck Manufacturing Co, E-Solutions

16. E. Steve Baughn, "Energy Efficiency in Aluminum Furnaces Using High Velocity Burners", Southwire case study

Energy Technology 2010
Edited by: Neale R. Neelameggham, Ramana G. Reddy, Cynthia K. Belt, Ann M. Hagni and Subodh Das
TMS (The Minerals, Metals & Materials Society), 2010

ENERGY SAVING IN THE FOUNDRY INDUSTRY BY USING THE "CRIMSON" SINGLE SHOT UP-CASTING PROCESS

Mark Jolly

School of Mechanical Engineering, The University of Birmingham
Edgbaston, Birmingham, B15 2TT, UK

Keywords: Foundry, Aluminium, Energy Saving, Rapid Melting, Life Cycle Analysis

Abstract

Instead of using the traditional batch casting process, the CRIMSON [1] (Constrained Rapid Induction Melting Single Shot Up-Casting) method employs a high-powered furnace to melt just enough metal to fill a single mould in a closed crucible. The crucible is transferred to a station for computer-controlled counter gravity filling of the mould for optimum filling and solidification. The CRIMSON method therefore holds the liquid aluminium for a minimum of time drastically reducing the energy losses attributed to holding the metal at temperature. With the rapid melting times achieved, of the order of minutes, there isn't a long time at temperature for hydrogen to be absorbed or for thick layers of oxide to form. The metal is never allowed to fall under gravity and therefore any oxide formed is not entrained within the liquid. Thus higher quality castings are produced leading to a reduction in scrap rate and reduced overall energy losses.

Introduction

Engineers at the University of Birmingham, England, with local company, NTec, have invented a new casting process that will reduce the energy costs of light-metal foundries. The technology, entitled CRIMSON, means that foundries need only heat the quantity of metal required to fill a single mould rather than large batches that use unnecessary energy and create waste. The complete concept of CRIMSON requires a change of philosophy in the approach to casting and requires the user to think holistically in order to reap the full benefits from the process. This paper will discuss the various aspects of energy wastage within the foundry sector and discusses the processes whereby the energy can be reduced. Primarily the author focuses on the aluminium sector but CRIMSON can be applied to many of the non-ferrous "light" alloys. Ferrous materials and non-ferrous "heavy" alloys can also benefit from some of the ideas but not necessarily directly from the application of CRIMSON.

The Current Situation

The philosophy that has been used in the foundry sector for many hundreds of years has been that it is more efficient to melt large volumes of metal than to melt only the exact amount required for each mould at the time when you need it. This has led to large melting furnaces and holding furnaces albeit becoming more and more energetically efficient over the years but still essentially wasting large amounts of energy. Recent research by Tharumarajah [2] has investigated the mass flows within aluminium high pressure die casting processes including looking at the impact that different levels of recycled scrap have on CO_2 content of the process.

His work investigates electrical and gas inputs, as well as water flows and waste flows and process losses. However his methodology does not come up with an energy burden per mass of good castings shipped. Such figures have been produced in the past by analyzing the usage of energy in the foundry [3], however this ignores the energy burden of the incoming materials. The analysis in this paper attempts to put some figures against the processes used.

A typical light-metal foundry usually melts anything from 100 kg to several 10s of tonnes of alloy in a range of furnace types. The liquid metal will then be held at a superheat of about 100 °C (about 700 °C for aluminium) in a second "holding furnace", before transferring it into a ladle and finally pouring into the casting mould. This may be a batch or semi-continuous process. The charge metal for the furnace comes from a number of sources. It can be high energy content virgin aluminium in the form of foundry ingot or direct chill (DC) cast billets. It can also be so-called "secondary" metal with an energy content corresponding to only 5% of virgin stock. This is metal which has been recycled from scrap and has been refined using similar processes to virgin alloys. The third source of material for a foundry is its own in-house recycled material arising from manufacturing process of the casting and from in-house machining processes, if they exist. Most foundries have in-house guidelines as to how much in-house scrap they recycle. Some have a policy of sending the material out for toll re-melting at a secondary alloy plant.

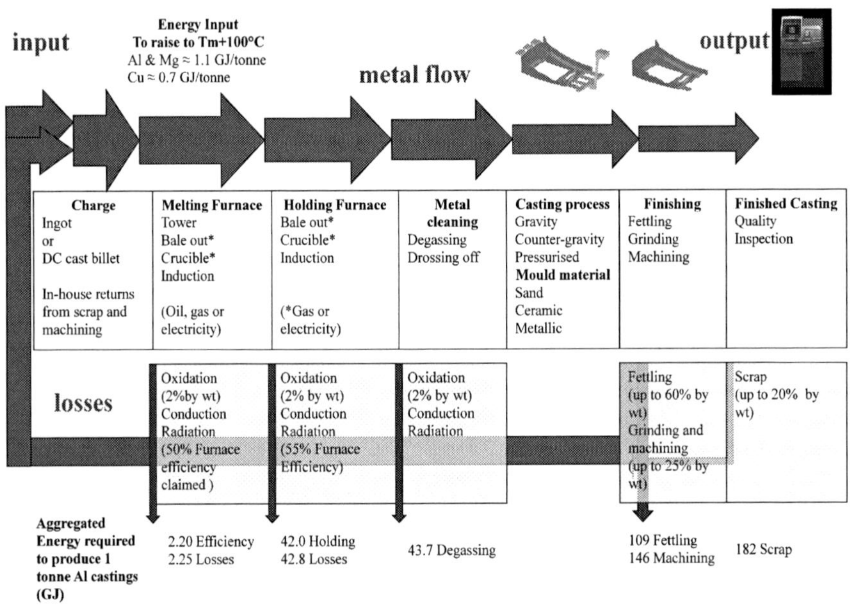

Figure 1: Flow route of aluminium through a typical foundry showing some typical losses and the initial estimated associated energy costs. Values for the melting losses and furnace efficiencies have been taken from industry literature [4, 5, 6].

It can take up to 8 hours for all the liquid metal in a batch to be used. Any leftover unused metal is poured off to be used again in-house, or scrapped to go for re-melting and refining in a secondary reprocessing plant.

Quality of Aluminium

Quality issues arise with the absorption of hydrogen from water in the atmosphere. Degassing by gas purging, or degassing additives on smaller quantities of metal, prolongs the time at temperature. Dross, a mixed metal oxide combined with small amounts of other compounds and some metallic aluminium, is created and scraped off, and either sent to landfill or, sometimes, for metal recovery. Added to this is the reaction of the metal with oxygen and water from the atmosphere, creating an oxide surface layer at every stage of the process. During pouring, this becomes mixed in, resulting in tiny particles or layers of metal oxide, reducing the quality of the liquid metal and eventually leading to reduced mechanical properties of the solidified castings [7, 8]. Furthermore, if the alloy specification is changed between batches, the furnaces and ladle are required to be flushed with more molten metal of each type before a new batch of the new metal type is used for production.

Melting and Holding

The traditional casting method uses energy to melt and subsequently keep a batch of metal at temperature (holding furnace), heat the pouring ladle, manage the problems associated with having kept molten metal hot (i.e. degassing and oxide formation), and melt unused metal, which is then recycled and/or scrapped, depending on quality. Figure 1 is a schematic of the foundry process and indicates the losses at each stage with the associated energy penalty for each stage, resulting in an overall energy cost per tonne of aluminium castings produced. The final figure of a massive 182 GJtonne^{-1} considers the worst case for every stage and indicates the potential for energy savings within this industry. This was calculated forward by taking the theoretical energy of melting aluminum. Figures published in 1997 [3] suggest that on average aluminium in the UK is produced using approximately 40 GJtonne^{-1}. In high-pressure die-casting foundries this ranged from 26 to 52 GJtonne^{-1}. For sand foundries the range was from 30 to 130 GJtonne-1. However, although these seem quite wide variations aerospace investment casting foundries are not featured and the embedded energy of the materials is not considered. Embedded energy being defined as the energy that energy used to produce the raw materials, in this case the virgin and recycled aluminium charge, the energy for producing resins and steel for dies for example. The energy for producing aluminium from ore, that is the embedded energy within virgin aluminium, is of the order of 55 GJtonne^{-1}[9].

The following sections deal with the various factors contributing to the huge amount of embedded energy contained within the casting produced

Melting. The theoretical quantity of energy required to raise 1 tonne of pure aluminium to its melting point, melt it and raise the temperature to a superheat of 100 °C can be calculated as 1.1 GJ. The major alloying elements of Si, Cu and Mg seldom total more than 15% by mass of most alloys. Table I gives some published figures of heat capacity and latent heats of fusion for a number of alloys showing that there is little difference in the calculated total energy to raise the temperature and hold at a suitable superheat for casting. In fact looking across most of the engineering metals with the exceptions of Cu and Zn the energy requirement to take the metal to

100 °C superheat is always approximately 1 GJ [6].Most furnaces achieve efficiencies between 50 and 60% and often the industry assumes the conservative figure of 50% [5 (p119)].

These data are confirmed by investigating furnace manufacturer's specifications [4]. A 1 tonne crucible furnace rated at 126 kW can melt at a rate of 295 $kghr^{-1}$. Published figures state that they can achieve 80% efficiency if a lid is employed, thus the melt rate becomes 236 $kghr^{-1}$. Using these figures the energy to melt 1 tonne can be calculated as 1.92 GJ. Thus for a tonne of pure aluminium the efficiency is about 56% and for the alloys ranging from 50 to 52%.

Many foundries will use tower furnaces as they produce a constant supply of liquid metal. A typical example may be a tower furnace rated at 6.6 $tonnes.hr^{-1}$ with an efficiency of 50%. 2.2 $GJtonne^{-1}$ would be expected to be used to melt however there are also losses in oxidation and volatilization of certain elements which it is estimated make up about 2% by mass of any

Table I. Energy to raise 1 tonne of Al alloy to a superheat of 100 °C

Alloy Designation US/UK	**Major Alloying elements**	**Solidus (T_s)**	**Liquidus (T_l)**	**Heat Capacity @ 25 °C**	**Latent heat of fusion @ melting point**	**Calculated total energy to raise 1Mg to T_l + 100 °C**
		°C	**°C**	**$Jkg^{-1}K^{-1}$**	**$kJkg^{-1}$**	**MJ**
Pure Al	NA	660	660	917	397	1071
356/LM25	7%Si 0.3%Mg	555	615	963	389	996
336/LM 13	11%Si 1%Cu 1.2%Mg 1%Fe 0.5%Mn	540	565	963	389	981
319/LM4	5%Si 3%Cu 0.8%Fe	515	605	963	389	957

Data from [10, 11, 12]

melt. Therefore producing 1 tonne of good aluminium requires an estimated 2.245 GJ of energy.

<u>Holding.</u> Melting large volumes of metal inevitably requires it then to be held. Foundries often use more energy for holding than for melting [3]. Making the assumption that a shift is 8 hours then every tonne melted in a shift is held on average for 4 hours if all the metal is used over that period. Holding usually has similar losses to melting which are estimated at about 2% per unit mass [5] and will be affected by things such as the amount of stirring, the atmosphere and whether the melt is covered. Alloy composition will also affect the losses experienced. It would seem logical to assume losses are increased per unit mass the longer the melt is held for however it is not easy to find any good data in the literature that can be used to estimate melt losses as a function of holding time. It has been proposed that holding as little metal as possible for as short a time as possible at the lowest possible temperature is the best solution. Some have even proposed that metal should not be held at all and used as soon as melted [p21,3]. The current

research programme at Birmingham University will hopefully provide some data on metal holding.

Degassing and Cleaning. Degassing of aluminium has become an essential part of most foundry practice. The majority use rotary degassing as an efficient and environmentally friendly method. The losses can be as high as 5% in terms of mass. Energy is also consumed by the rotor and there is embedded energy in the nitrogen or argon gas used as it will have been compressed into the cylinders. A midrange rotary degassing unit might be powered by a 3.5 kW motor and be used for 15 min. The amount of gas used each campaign would at 20 L min^{-1} would be 300 L. Thus the energy of N_2 or Ar embedded would be about 0.5 MJ. Combined with the energy for the rotor this means each campaign has an embedded energy of approximately 2 MJ. This might be carried out 4 times during a shift giving 8 MJ.

Cleaning usually refers to the removal of oxides. This is accomplished by either incorporating the process with the degassing by injecting a powdered flux though the lance or rotor, or by fluxing, rabbling the surface and "drossing-off". The production of dross results in melt losses and there is again inevitably embedded energy in the cleaning products used. The embedded energy in flux products is something that has not been investigated as far as the author is aware.

Assuming flux is added at a rate of 0.5% per tonne of aluminium and assuming the flux is a K or Na fluoride then the energy for melting is 0.971 MJ/kg and 1.6 MJ/kg [12] respectively and heating and melt losses of 5% equates to 5.1 and 8.8 MJ/tonne of aluminium, respectively.

Mould, Die and Core making

All foundries require a cavity for the liquid metal to be poured into. These can be made from a variety of different materials all with different energy burdens of their own. For consumable mould processes there will be the energy burden in the sand or ceramic materials from the mining and the transportation. These materials also use binders, organic or inorganic each with an energy content. Sand moulds could be formed with a wooden or steel pattern each with a very different energy. Ceramic investment shells are formed around a polymer material which has been injected into a steel die and then the polymer is removed using steam. The energy content of this process could potentially be extremely high. For non-consumable processes like die-casting the steel or cast iron dies have an energy content that would be of the order of those published by Brimacombe et al. [13] for steel products.

Finishing processes

Finishing processes have a massive effect on the overall process efficiency.

Fettling. It is at this point that in the process that the so-called "method" (rigging in the US) is removed. These are the extra bits of metal that enable the casting to be manufactured i.e. the feeders, risers and running system. In a good foundry producing general engineering or automotive castings a figure of about 50% may be achieved. In a high precision aerospace investment foundry this figure could rise to 90%, implying that only 10% of the metal cast ends up in the black casting.

Grinding and machining. In order to extract the casting for the die or moulds, there is often a taper or draft angle added to the final component shape. This may or may not need to be machined or ground off. Any such process uses energy thus the less finishing machining that is carried out the lower the energy content of the final component. The energy involved in machining and fettling will be measured in the new research programme at Birmingham.

Scrap. Any component that contains porosity in locations that will affect the component performance, or that have unacceptable surface finish, or contain inclusions such as oxides, will be scrapped. These materials will go back to the melting stage, to the secondary re-melter or toll re-melted in any case the embedded energy is everything that had been carried out on the casting at the stage. Estimates of the level of embedded energy are considered in the section on Life Cycle Analysis.

The Crimson Route

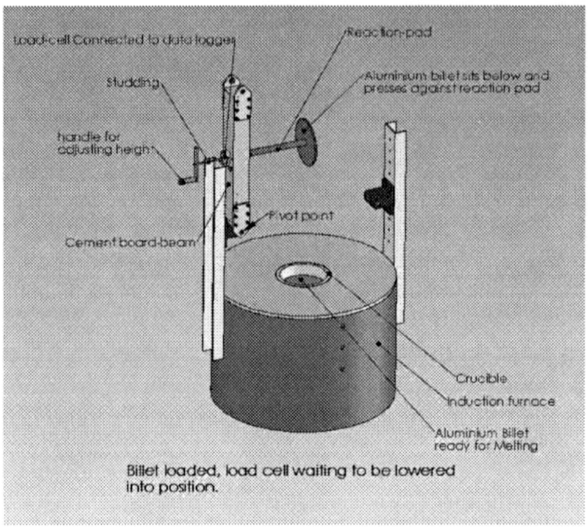

Figure 2: Image of the CRIM part of CRIMSON process

Figure 3: Photograph of the up-casting unit at Birmingham University

The Crimson route for casting should enable foundries to save energy at various process stages. Aluminium will be provided in a pre-alloyed, clean and degassed condition from the metal supplier. The metal is heated rapidly with an induction coil and used immediately. This will obviate all the holding and degassing losses and will reduce melting losses. Induction melting is also a much more efficient melting method than either radiant, gas or oil fired furnaces.

The equipment at the University of Birmingham comprises a 350 KW induction furnace which is expected will melt up to 33 kg of aluminium alloy in less than 3 minutes. DC cast billets of the selected alloy are cut to weight and placed within the crucible. Power is turned on and the metal is constrained within the furnace to obviate the problem of ejection of the metal (Figure 2). When the metal is molten the crucible is transported to the up-casting unit (Figure 3). The crucible is quickly indexed into the casting position as soon as the safety inter-locked doors are closed. The software controlling the motor is then activated automatically and applies the correct displacement cycle to the screwed injector of the up-caster. The base of the crucible moves up through the crucible and is sealed against leaking by a gasket.

Life Cycle Analysis (LCA)

There are a number of different methodologies for dealing with Life Cycle Analysis and sometimes known as Life Cycle Inventory analysis (LCI). For the aluminium foundry sector it is probably best to use those that have been used in the steel industry and adapt them for the differences [13]. Within the foundry sector the amount of recycling and levels of scrap that exist are highly dependent on the product and end market. Thus in general, the yield and scrap rates in the aerospace sector are respectively low and high, whereas in the automotive sector these tend be higher and lower than in the aerospace sector. Both of these factors have an effect on the total effective recycling efficiency of the process and thus the effect of the burden of the energy of the virgin and recycled aluminium used. There are a number of different ways carrying out an LCA but in this paper the focus is on two methods. One is the Multiple Recycling Concept and the other is the Closed Loop recycling as described by ISO14040 [14]

.

Boundaries and Process Measures

One of the most important things to decide on when carrying out an LCA are what are the process boundaries. In the study by Tharumarajah [2] the boundaries are limited to the inflow of aluminium, gas, electricity, water and die lubricant coming on site and aluminium, water and waste leaving the site. However the intrinsic embedded energy content of these material flows is not considered. The study only considers the 1st pass material efficiency which is reported as 93.9%, waste is reported as 6.1%. This is equivalent to the process yield (Y) discussed by Brimacombe et al. [13]. However using an Operational Material Efficiency (OME) also known as True Yield [3] i.e. the percentage of good metal shipped as a proportion of metal melted, the picture is changed somewhat, as this is 47.7%. This operational material efficiency is a good measure of the quality of the process and even though in this scrap material is recycled internally the level of scrap is high and thus the OME is poor. In this paper the embedded energy of the materials input is also considered and the effect of multiple recycling calculations is discussed. However the area that neither Tharumarajah nor this paper studies in detail is the energy burden of the mould, die or core materials. The current research programme at the University of Birmingham will hope to address this issue.

Multiple Recycling Method

In order to estimate the embedded energy using this method it is important to measure or estimate a number factors:

Process Yield (*Y*). This factor is a number, usually less than 1, that is calculated by taking away the true mass losses in the process from unity. In the case of the aluminium foundry the true mass losses are those from melting, holding and degassing as these are oxidation losses. The other "losses" can be recycled in the process. Using the estimates above the real mass losses come to a total of 8.8%. The process yield, Y, is therefore 0.912 or 91.2%.

Recovery Ratio (*RR*). This is the figure that considers the scrap that is recycled from the process as a percentage of the material put in. Taking a foundry in the automotive or general engineering sector it can be estimated that the worst case RR is about 0.64 (64%). For an aerospace foundry this is potentially higher as the proportion of good castings is much lower. It could be as high as 0.86 (86%) but very often in this sector only virgin aluminium is used and thus RR would be zero (0%).

Recycling Efficiency (*r*). This parameter represents how efficient the process is over one cycle. It is the product of the Process Yield and the Recovery Ratio. Thus

$$r = RR \times Y \qquad \text{Eq. 1}$$

Calculating LCI for complete systems. To calculate the LCA using Multiple Recycling the following equations are required.

The total mass passing through the chosen number of cycles, M

$$M = 1 + r + r^2 + ... + r^{n-1} \quad \text{Eq. 2}$$

The total energy content for the chosen number of cycles can be calculated using equation 3.

$$\text{Total energy content} = X_{pr} + rX_{re} + r^2 X_{re} + ... + r^{n-1} X_{re} \quad \text{Eq. 3}$$

Where X_{pr} is the energy form the primary process and X_{re} is the energy for the recycling process. Combining these two equations we can now get a Life Cycle Inventory (X) defined by equation 4.

$$\text{LCI for the whole system} \quad X = \frac{X_{pr} + rX_{re} + r^2 X_{re} + ... + r^{n-1} X_{re}}{1 + r + r^2 + ... + r^{n-1}} \quad \text{Eq. 4}$$

This can be reduced to equation 5.

$$\text{LCI for the whole system} \quad X = (X_{pr} - X_{re})\left[\frac{(1-r)}{(1-r^n)}\right] + X_{re} \quad \text{Eq. 5}$$

The energy for each cycle can be plotted and it can be seen that the curve is asymptotic to a value that can be given by equation 6 as $n \to \infty$ and $r^n \to 0$ and $(1 - r^n) \to 1$.

$$X = X_{pr} + r(X_{re} - X_{pr}) \quad \text{Eq. 6}$$

Suggested values of Y, RR, X_{pr}, X_{re} and r for aluminium foundries in the automotive/general engineering market, and aerospace sector, with and without recycling, are given in Table II. Figure 4 is a plot of the energy burden for each cycle until it reaches the asymptote.

Table II: Suggested LCI parameter values for aluminium foundry sectors

Sector	**OME**	**Process Yield, Y**	**Recovery Ratio, *RR***	**Recycling Efficiency, *r***	**LCI value for primary process X_{pr} (GJtonne^{-1})**	**LCI value for recycling process X_{re} (GJtonne^{-1})**
Automotive/ General	27%	64%	91%	0.583	55.080	2.754
Aerospace no recycling	5%	0%	91%	0.000	55.080	2.754
Aerospace with recycling	5%	86%	91%	0.782	55.080	2.754
CRIMSON route	34%	66%	99.5%	0.656	55.080	2.754

Closed Loop Recycling

Closed loop recycling is easier to apply and essentially can be applied where there are no inherent changes to the materials. This is the methodology recommended in such cases in ISO 14041. This can therefore be used for aluminium castings. The same result is obtained and it is a much simpler to apply. There are two stages for the calculation. The first stage of the calculation is to estimate the energy burden of the scrap. The method is clearly shown by Brimacombe et al. [13**Error! Bookmark not defined.**] for steel. The second is to carry out the LCA for the recycling route. These calculations are shown below for a general/automotive type foundry and for an aerospace foundry.

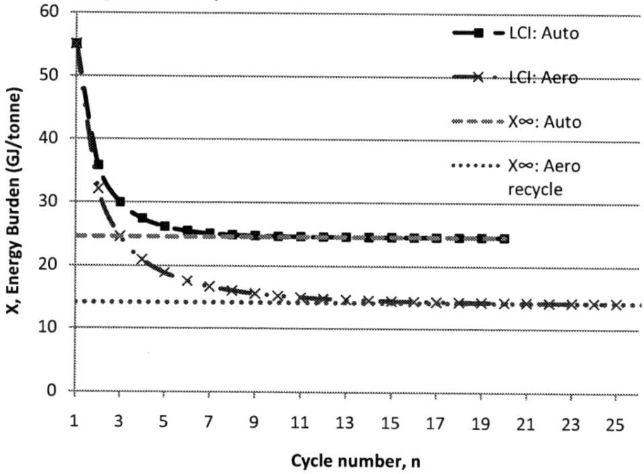

Figure 4: Energy burden per cycle for aluminium castings in various market sector foundries

Automotive sector cradle to grave calculations

Primary Route	Bauxite →	**55.1 MJ**	1 kg Aluminium	→	Casting, fettling, machining	→	0.64 kg scrap
Energy for re-melting		**2.8 MJ**	Re-melt losses		0.088 kg		
Secondary Route	1.096 kg →	3.0 MJ	1 kg Aluminium	→	Casting, fettling machining	→	0.64 kg scrap
Energy value of scrap			$(55.1 - 2.8)/1.096 = 47.7\ MJ/kg$				
Primary route	*Net scrap output 0.64 kg; equivalent to energy credit of 47.7 x 0.64 =* **30.5 MJ**						
Energy of Primary route	**55.1 MJ** Process impact	less	**30.5 MJ** Energy credit for scrap			→	**24.6 MJ/kg**
Secondary route							
	Net scrap input of 1.096-0.64 = 0.46 kg; equivalent to an energy input of 47.7 x 0.46 = **21.8 MJ**						
Energy of Secondary route	**21.8 MJ** Scrap input	plus	**2.8 MJ** Process impact			→	**24.6 MJ/kg**

Discussion

The melting of metals and subsequent use of the liquid in manufacturing shapes has enabled man to develop some incredible technologies including the silicon chip and the single crystal turbine blade which is now ubiquitous in the modern jet engine. The down side of this old but enormously valuable technology is the large amount of energy required to achieve this outcome. In the current era of sustainability and reduction of CO_2 emissions that have accelerated climate changes on the planet it is incumbent upon any such energy-hungry industry to identify how it can respond to the situation by reducing energy usage and being efficient with the energy that it uses.

Aerospace sector cradle to grave calculations

Primary Route	Bauxite →	**55.1 MJ**	1 kg Aluminium	→	Casting, fettling, machining	→	0.86 kg scrap
Energy for re-melting		**2.8 MJ**	Re-melt losses		0.088 kg		
Secondary Route	1.096 kg →	3.0 MJ	1 kg Aluminium	→	Casting fettling machining	→	0.86 kg scrap
Energy value of scrap		$(55.1 - 2.8)/1.096 = 47.7\ MJ/kg$					
Primary route	*Net scrap output 0.86 kg; equivalent to energy credit of 47.7 x 0.86* = **41.0 MJ**						
Energy of Primary route		**55.1 MJ** Process impact	less		**41.0 MJ** Energy credit for scrap	→	**14.1 MJ/kg**
Secondary route	*Net scrap input of 1.096-0.86 = 0.24 kg; equivalent to an energy input of 47.7 x 0.24* = **11.3 MJ**						
Energy of Secondary route		**11.3 MJ** Scrap input	plus		**2.8 MJ** Process impact	→	**14.1 MJ/kg**

CRIMSON cradle to grave calculations for automotive

Primary Route	Bauxite →	**55.1 MJ**	1 kg Aluminium	→	Casting, fettling, machining	→	0.66 kg scrap
Energy for re-melting		**2.8 MJ**	Re-melt losses		0.005 kg		
Secondary Route	1.005 kg →	2.8 MJ	1 kg Aluminium	→	Casting fettling machining	→	0.66 kg scrap
Energy value of scrap		$(55.1 - 2.8)/1.005 = 52.1\ MJ/kg$					
Primary route	*Net scrap output 0.66 kg; equivalent to energy credit of 52.1 x 0.66* = **34.3 MJ**						
Energy of Primary route		**55.1 MJ** Process impact	less		**34.3 MJ** Energy credit for scrap	→	**20.8 MJ/kg**
Secondary route	*Net scrap input of 1.005-0.66 = 0.35 kg; equivalent to an energy input of 52.1 x 0.35* = **18.0 MJ**						
Energy of Secondary route		**18.0 MJ** Scrap input	plus		**2.8 MJ** Process impact	→	**20.8 MJ/kg**

In this paper Life Cycle Analysis (LCA) using both infinite recycling and closed loop methods has been applied to try to understand the energy burdens that are carried by aluminium castings in the automotive and aerospace sectors. The author is assuming that for the aerospace sector the majority of casting are produced using the investment casting process whereas for the other sectors it will be a mixture of mainly high and low pressure die and sand casting.

LCA is notoriously complicated to carry out because of the difficulties in the definition of the process boundaries. An example of this can be seen in the aerospace sector. Generally, the aerospace sector uses virgin aluminium as feedstock for the casting process. There is often a high scrap rate which can be higher than 90%. This is not all scrap from bad product it is made up of process scrap which comes from the methoding (rigging) and machining. This becomes waste in the aerospace sector but is a high quality, high energy content raw material for the rest of the aluminium foundry sectors. The question then is should that be included in the aerospace LCA to enable a complete picture to be obtained or excluded and just added into the other sectors? Looking at the energy burden estimated in each of cases from the LCA can be misleading unless one understands what this is representing. For an aerospace sector casting with no internal recycling the energy burden per kg of melted metal is approximately 55 MJ. However this kg of liquid metal may sometimes only produce 55 g of good castings! Thus the burden per kg of good castings could be as high as 1 GJ. The scrap which is estimated to have a burden of about 48 $MJkg^{-1}$ is then fed back into the other aluminium sectors. If it was fed back into the aerospace sector and used as part of the charge material then *because* of the high scrap rate the energy burden would appear to be reduced considerably to about 14 $MJkg^{-1}$ as less primary aluminium would be used. This is a reflection of the very considerably less energy required to re-melt aluminum compared with the energy required to electrolytically reduce it from its ore. However this may still only produce 55g of good castings so, although reduced, the energy burden per kg of good castings is still an enormous 258 MJ. It becomes very obvious from these calculations that the energy burden is heavily affected by the recycling efficiency (r) *and* the operational material efficiency (OME). So although the aerospace sector *could* have a recycling efficiency of 78% because in practice the OME is so poor the energy burden per unit mass of good castings is very high. This information can be used by the industry in order to manage change within the sector.

Taking a similar approach to the traditional methods for producing automotive castings it can be seen that, as the scrap rate is lower, the energy credit is not as high, so for each kg of liquid metal the energy burden is actually higher than in the aerospace sector at about 25 $MJkg^{-1}$ which is still less than 50% of the primary aluminium route. Using the same approach as in the previous paragraph and incorporating the OME it can be seen that for this sector an average energy burden might be of the order of 90 $MJkg^{-1}$. This figure is surprisingly similar to the energy measurements made in the 1990s and published by the UK government [3]. The energy levels published are for the energy *used* in the foundries so do not reflect energy burden of the materials coming into the process. Thus the only part considered in the calculation in this paper is the impact of the secondary melting at 2.8 $MJkg^{-1}$ melted which is equivalent to 10 MJ for each kg of good castings. It could be proposed therefore that to the figures measuring the use of energy by the foundries a further 80 $MJkg^{-1}$ (80 $GJtonne^{-1}$) could be added to truly represent the impact that aluminium castings have on the environment.

So what effect can the CRIMSON process have on the foundry sector when it is apparent that effective methoding to improve the OME is one of the major ways of reducing the energy content as long as recycled material from end of life castings is used as the raw materials rather than virgin aluminium? An LCA for the CRIMSON process keeping everything equal except for the effect on the melting shows that the energy burden 63 $MJkg^{-1}$ with an OME of 30%. This is a considerable reduction of the 90$MJkg^{-1}$ for the traditional processes. With the improvements in scrap rates expected, as the material will be cleaner, but with no change of methoding then burden reduces by very little to 62 $MJkg^{-1}$ and an OME of 34%. Improvements to the OME by altering the methoding could change this dramatically and this may be where the biggest second opportunity arises. For example reducing the methoding by 50% (i.e. improving the so-called box yield to 67%) could bring the OME to close to 50%. This has the effect of reducing the embedded energy per kg of good castings to 59 MJ. These calculations assume that only supply of aluminium is the internal recycled scrap and primary aluminium. If the internal scrap is topped up with end of life scrap from other sources and the primary aluminium is kept to a level of about 30% then the embedded energy drops dramatically to about 40 $MJkg^{-1}$. Obviously traditional process can also raise their level of recycled materials to improve the situation with regard to embedded energy but the effect of the CRIMSON process in dropping the energy to 70% of that traditional route by reducing melt losses and obviating holding and degassing losses is dramatic.

Conclusions

High scrap rates and low operational material efficiency in the aerospace sector leads to an extremely high energy burden per tonne of good castings estimated to be in the region of 1 TJ.

If the aerospace sector were to recycle its own scrap internally the energy burden per tonne of good casting could be reduced by 75% to approximately 250 GJ based on the assumptions used I this paper.

In the general and automotive aluminum foundry sectors the energy burden is estimated to be of the order of 90 GJ per tonne.

Applying the CRIMSON process to the general and automotive sectors has the effect of reducing the melting losses and obviating holding and degassing losses. This has a positive effect on the energy burden by reducing it to about 63 GJ per tonne of good castings.

Keeping primary aluminium content to 30% and improving the running systems could reduce this energy burden to a level of 40 GJ per tonne of good castings shipped.

Acknowledgments

This project is funded by the Engineering and Physical Sciences Research Council (EPSRC) of the UK under grant number EP/G060096/1.

References

1 Patent Pending P306306GB
2 A Tharumarajah, "Benchmarking aluminium die casting operations", Resources, Conservation and Recycling 52 (2008) 1185–1189
3 Energy Consumption Guide 38, "Non-Ferrous Foundries", 2nd Edition, Best Practice Programme, Department of the Environment, Transport and the Regions, October1997
4 Nabertherm – www.nabertherm.com
5 Foseco Non-Ferrous Foundryman's Handbook, Ed. John Brown, Pub Butterworth Heinemann, Oxford, 1999
6 MR Jolly, "Castings", Comprehensive Structural Integrity, Elsevier, Oxford, UK, Eds. I. Milne, R.O Ritchie and B. Karihaloo, ISBN 0-08 044157-2 and ISBN 0-08-043749-4, 2003, Chapter 18, pp 377 - 466
7 NR Green & J Campbell, "Statistical distributions of fracture strengths of cast Al-7Si-Mg alloy", Mat. Sci. and Engineering, 1993, A173, pp. 261-266
8 J Campbell, "Invisible macrodefects in castings", Journal de Physique IV, Coloque C7, supplément au Journal de Physique III, 3, Novembre 1993
9 http://www.world-aluminium.org
10 CJ Smithells (ed.), 1976, "Metals Reference Book", 5th edn., Butterworths, London and Boston.
11 ASM Metals Handbook, 9th Ed., ASM, Metals Park Ohio, USA, 1979
12 DR Lide, CRC Handbook of Chemistry and Physics, 74th Ed., CRC Press, Boca Raton Florida, US, 4-108
13 L Brimacombe, N Coleman and C Honess, Recycling, reuse and the sustainability of steel, Millenium Steel, 2005, p31-36
14 ISO 14040, Environmental management, "Life cycle assessment: Principles and framework", International Organisation for Standardisation (ISO), Geneva, 2006

Energy Technology 2010
Edited by: Neale R. Neelameggham, Ramana G. Reddy, Cynthia K. Belt, Ann M. Hagni and Subodh Das
TMS (The Minerals, Metals & Materials Society), 2010

Energy and Emissions with Oxyfuel

Thomas Niehoff[1]

[1]Linde Gas, 85716 Unterschleissheim, Carl-von-Linde-Str. 25, Germany

Keywords: Energy, Oxy-fuel, Furnaces, Combustion Systems, Emissions

Abstract

Oxyfuel combustion systems are state of the art for many melting operations. Airfuel systems have been stepwise modified to oxygen enrichment, oxygen lancing and oxyfuel. Every change of the melting process will have consequences with regards to emissions and energy usage. The uniqueness of many operations requires unique skills to resolve potential issues. This paper will give some examples of how airfuel systems can successfully be converted to oxyfuel.

Introduction

Oxyfuel technology – meaning the combustion of fossil fuels such as natural gas, oil or coal with technical oxygen instead of air for industrial melting or heating processes is widely used in many industries these days. Even though oxygen production, storage and handling has been state of the art for decades other opinions exist. Often oxyfuel is regarded with fear and potential hazards like: “Oxygen makes the process too hot.”, “Oxygen will melt away the refractory.”, Liquid oxygen is dangerous.”, “Oxygen will cause explosions.”, ……etc.
Typically a more detailed discussion about a specific industrial process and how oxygen can be applied in a safe and economic way the seeming issues go away.

Oxygen Conversion (stepwise)

The development from airfuel based heating processes to oxyfuel based heating processes can go the following path (see Fig. 1).

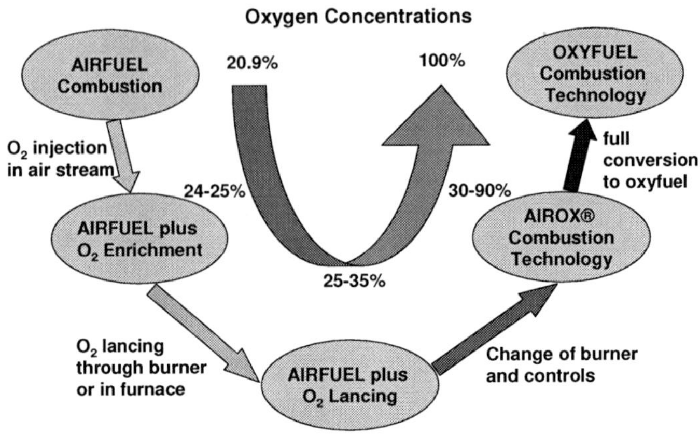

Figure 1. Development from airfuel to oxyfuel.

Air contains about 20.9 vol.-% of oxygen. When technical oxygen is mixed with air the oxygen concentration goes up. This is called oxygen enrichment. When no air or nitrogen is present in the oxidizing gas then the process is called oxyfuel (with pure technical oxygen). Over the time LINDE has developed various technologies how to use oxygen in combustion processes in a safe, reliable and economic way. Table 1 gives an overview.

LINDE Technologies	Description
AIROX® Combustion Technologies	Air – Oxy – Fuel Combustion
FLEXFLAME® Combustion Technologies	Adjust the flame length
WASTOX® Combustion Process	Burn charged hydrocarbons
Low Temperature Oxyfuel Combustion Technologies	Adjust the flame radiation
HIGHJET® TDI Oxygen Technology	Shaft furnace/Cupola optimization

Table 1: Overview of LINDE combustion technologies/processes.

Basic Furnace Conditions

All processes that may be further optimized need to be understood in detail. When changing burners or process control to a melting operation a lot of risks are being taken. Analyzing the current operation as a base line operation is the first step. For this the whole process is being monitored and analyzed for the current situation. For example readings from fuel and air volumes are taken to determine the specific energy consumption per ton of metal produced. Flue gas analysis of the main components like CO_2, CO and O_2 give a good indication of the efficiency of the combustion process inside the furnace and helps to quantify air leak volumes.

Cycle times (charging, melting, holding and tapping) are taken and processed. This will lead to melt rates and energy requirements for each specific interval of the melting process. The way of charging and firing during charging is very important for the rest of the process intervals. Charging can be done is several steps and various furnace and firing conditions. How much is being charged? When? What quality material is being charged?

The fact that solid metals have a higher thermal conductivity than liquid is not often utilized. Solid metal has approximately three times the heat conductivity of liquid metal. That means that solid metal can heat up about three times faster compared to liquid metal. Modern combustion technologies like Low Temperature Oxyfuel Burner and AIROX® Combustion Technologies from LINDE can vary between convective and radiant heat transfer. How is the material being charged? Is it in the shadow of the flame or does it see the bright and radiant flame? Is the material being overheated or evenly heated and melted? All this can be affected by choosing the right combustion system and process conditions.

Thermal Heat Conductivity [W/ m * K]				
Metal	0 °C	200°C	400°C	Liquid Metal
Copper (Cu)	378	372	363	
Aluminum (Al)	227	227	227	87
Lead (Pb)	35	33		14

Table 2: Thermal heat conductivity from various metals.

Dross generation is the result of burning aluminum during the melting process. This oxidation is mainly caused by local overheating and having oxygen available to oxidize the metal. Aluminum takes the oxygen from free oxygen molecules H_2O, CO_2 and even CO. The high affinity to oxygen from aluminum is causing this. Dross generation can be measured and optimized. The fact that pure oxygen may be used for combustion does not necessarily result in higher dross formation. Even the opposite is the case. Improved burner and process control with oxyfuel will allow higher productivity at reduced cycle times which typically results in improved aluminum yields. The dross generation of the base case operation needs to be understood before modifying the process or converting to oxygen. This is the basis for recovery improvements.

How does the heat get to the charged material? In Figure 2 the main heat transfer mechanisms are shown.

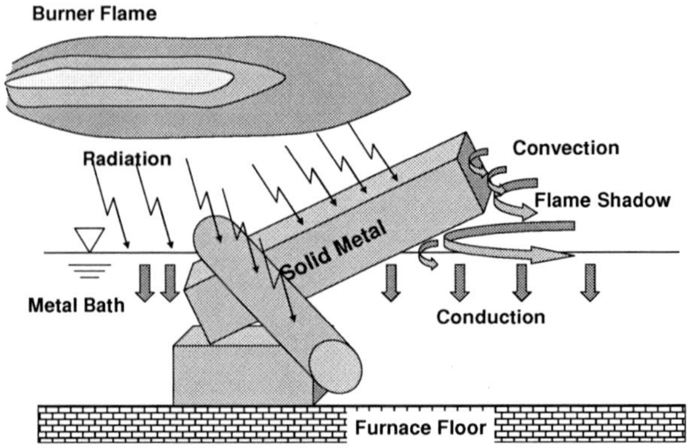

Figure 2: Heat Transfer to charged material.

Basic Firing Conditions

A bright flame will radiate heat in all directions. The radiation will go straight and take no turns or bends and not go behind covered areas. The flame will also produce hot gases called combustion products. These hot gases follow a typical flow pattern in a specific furnace. Hot gases can flow behind shielded areas and thus reach sections of the charge material where radiation cannot reach. For solid charges where the material is piled up in the furnace in a way that the flame creates a lot of shadow areas it is very likely that the heating of the metal will result in local overheating directly at the flame and cold spots in the shadow areas.

Dross generation will be high in the overheated spots and heating/melting will be low in the shadow areas. Due to different heat conductivities with regard to their state (liquid or solid) also adjustments to the burner firing rate are required. In many melting operations typically during charging with open door the burners are shut off and after charging the burners run on high fire until tapping. To operate this way is in many ways not optimized: in terms of energy efficiency, process requirements and cost. To take advantage of the given physical conditions and the specifics from charging process, into melting and into tapping we suggest to run a programmed melt cycle where the firing modes of the combustion system only run in optimum conditions. That means that overheating is greatly avoided and firing is hard when required only adding the necessary energy to the process. The LINDE technology portfolio even allows to adjust the firing mode from bright flame to flameless according to process requirements. The Programmed Melting will enhance the melting result in terms of energy, yield and cost.

Energy and Emissions

The effect on energy and emissions has been discussed in [1], [2] and [3]. Energy savings of 50% and more can be realized when switching from airfuel to oxyfuel. The nitrogen content in the air is close to 80%. In airfuel combustion the inert nitrogen gets heated up from ambient to furnace temperature and then leaves the process with the flue gases. This takes up most of the energy. CO_2 emissions are directly related to the quantity of fuel needed and the composition thereof. For instance hydrogen (H_2) could be used as a fuel. The flue gas from hydrogen would mainly be water vapor (H_2O) and no CO_2. Hence the amount of carbon in the fuel is directly related to the CO_2 emissions. When the natural gas/ energy per ton of metals are halved then also the CO_2 emissions are halved, since they are directly related to the fuel usage.

Emissions of CO typically occur when a combustion in incomplete, i.e. too much fuel or not enough oxygen. Combustion processes use a ratio of fuel to oxidizer (air or oxygen) to ensure complete combustion and no CO emission. Oxyfuel increases the retention time of combustibles in the hot zones and the reactivity of fuel and oxidizer because of missing nitrogen and reduced gas volumes. CO emissions are reduced when converting from airfuel to oxyfuel.

NO_x is generated in flames where nitrogen and oxygen is present. At higher temperatures (above 1000°C/2000° F) the NO_x generation is higher due to thermal NO_x. Fuels can contain nitrogen. Nitrogen free fuel combusted with oxygen will burn without NO_x. Industrial furnace conditions are not ideal. Air leaks and nitrogen containing fuels are very common. Low NO_x burners are used to lower the NO_x emissions. Low NO_x burners stage the combustion process and reduce the flame temperature.

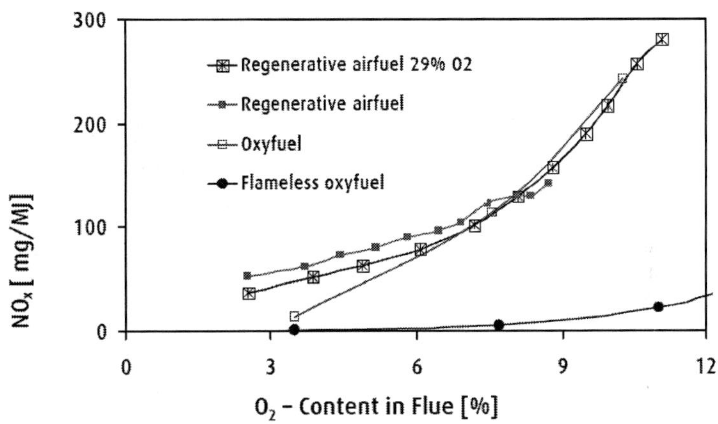

Figure 3: NO_x emissions of different combustion technologies.

NO_x generation is affected by many parameters. In Figure 3 the influences of different firing technologies is shown. The more oxygen is available in the flue gas, the more NO_x can be formed. Regenerative airfuel burners and conventional oxyfuel burners show similar NO_x numbers in Figure 3. The flameless oxyfuel technology from LINDE clearly generates less NO_x compared with the other burners. NO_x also depends upon furnace temperature and very specific furnace conditions, which will not be covered in this paper.

Conclusions

Oxyfuel is mainly used to reduce cost of metal production in various industries. Oxyfuel impacts the whole melting process and requires changes when starting from airfuel operation. Oxygen does reduce energy consumption and increases production. Side effects are reduced emissions like CO_2 and NO_x. A controlled way to optimize the melting operation is being introduced by LINDE. Programmed melt cycles improve consistency of results and ensure low cost operation.

References

1. T. Niehoff "Oxyfuel – Solutions for Energy and Environmental Conservation", TMS 2008, Feb. 2008, New Orleans, LA.
2. T. Niehoff "Oxyfuel – Energy efficient melting", TMS 2009, Feb. 2009, San Francisco, CA.
3. T. Niehoff, "Optimised Aluminium Melting", 10. OEA International Aluminium Recycling Congress, Berlin, Mar. 2009.

Energy Technology 2010
Edited by: Neale R. Neelameggham, Ramana G. Reddy, Cynthia K. Belt, Ann M. Hagni and Subodh Das
TMS (The Minerals, Metals & Materials Society), 2010

INDUSTRIAL APPLICATION EXPERIENCES WITH MICROPOROUS CALCIUM HEXALUMINATE INSULATING MATERIAL SLA-92

Dale Zacherl[1]; Rainer Kockegey-Lorenz[2]; Andreas Buhr[3]

[1]Almatis Inc, 501 West Park Road, Leetsdale, PA 15056, USA,
[2]Almatis GmbH, Giulinistraße 2, 67065 Ludwigshafen, Germany
[3]Almatis GmbH, Lyoner Strasse 9, 60528 Frankfurt, Germany

Keywords: SLA-92, Refractory, Insulating, Aggregate

Abstract

The microporous calcium hexaluminate insulating material SLA-92 has been introduced as an alternative to refractory ceramic fiber and other insulating refractory materials. Key properties associated with this product include: high chemical purity, long term high temperature stability up to 1500 °C, low thermal conductivity up to 1500 °C, and high thermal shock resistance. The properties of the new material and various applications in the steel and ceramic industries have been reported in previous papers. The current paper summarizes the industrial application experiences with innovative refractories based on SLA-92 and discusses potential new applications in various manufacturing as well as energy-related industries.

Introduction

SLA-92 was introduced to the refractories industry in 1998 by van Garsel, et al. [1] as a new insulating raw material for high temperature applications, followed by investigations [2] of long term temperature stability for this new lightweight aggregate up to 1500°C.

SLA-92 was developed as an innovative alternative to commonly used refractory ceramic fibers. It has unique properties regarding chemical purity and mineral composition and is a microporous, silica-free calcium hexaluminate (CA_6 – mineral name: hibonite).

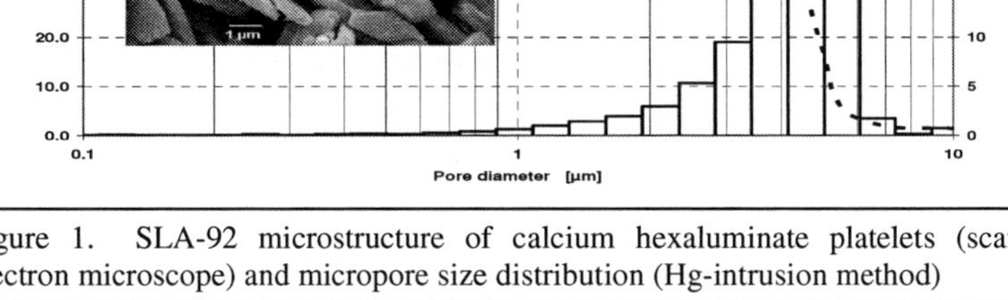

Figure 1. SLA-92 microstructure of calcium hexaluminate platelets (scanning electron microscope) and micropore size distribution (Hg-intrusion method)

SLA-92 chemical and physical data and mineral composition are listed in Table I, microstructure, open and relative porosity are depicted in Figure 1.

Table I. SLA-92 Product Data

Chemical Composition [wt.-%]	Typical	Min.	Max.
Al_2O_3	91	90	
CaO	8.5		9.2
Na_2O	0.40		0.5
SiO_2	0.07		0.2
Fe_2O_3	0.04		0.1
Physical Properties			
Loose Bulk Density [kg/l]	0.5 - 0.6		
Bulk Specific Gravity [g/cm³]	0.80		0.95
Mineral Phase Composition	CA_6 ($CaO \cdot 6\ Al_2O_3$) Major CA_2 Minor α - Al_2O_3 Minor		
Available Sizes	3 – 6 mm 1 – 3 mm 0 – 1 mm		

Measurements on experimental castables show excellent resistance to thermal shock. The low density in combination with the microporosity (which hampers heat transfer by radiation when temperatures exceed 1000°C) results in a low thermal conductivity of 0.4 W/mK at 1300°C, details are reported in [1].

This makes SLA-92 an innovative alternative to refractory ceramic fibers and it is an unmatched insulating aggregate for preferred use in state-of-the-art monolithic concepts, offering interesting application opportunities in the wide field of demanding high temperature insulation.

Since its introduction, numerous papers and publications report the use of SLA-92 as a key raw material for high temperature insulation materials in a variety of applications. On closer examination of the papers, topics are mainly focused on:

- Fiber replacement
- Innovative raw material for improved performance

Besides advantages resulting from chemical purity, physical properties, microporosity, long term high temperature stability and resistance to demanding thermal shock, obviously one main driver for use of SLA-92 is to have a feasible substitution of refractory ceramic fibers, which are classified as category 2 carcinogens by EU legislation. A German public sponsored research and development project [3] reports about substitution of fiber containing concepts by SLA-92 based insulating materials in high temperature processes in the steel making, ceramic and other industries. This paper summarizes the industrial application experiences with the new innovative refractories based on SLA-92.

Applications in the Steel Industry

Often the steel industry as main consumer of refractories is also the driver for refractory innovations. The focus for the last few years has been on the replacement of refractory ceramic fiber (RCF's), so far mainly for pre-heaters applied in steel making processes and the variety of reheating furnaces at hot rolling mills.

Ladle pre-heater cover lining

Duhamel and Verelle [4] reported about a new design of covers for ladle preheating in Dunkirk Steelworks. The objective was the elimination of drawbacks due to handling, installation and removal of refractory ceramic fibers, along with significantly increased service life and availability of the ladle covers. Emphasis is also on health aspects and additional cost involved by wrecking RCF linings and disposal of fiber wastes in compliance with the environmental legislation. A castable based on SLA-92 has been a key component of the solution to achieve the objective. Even at initial higher material cost of about 54% compared to the RCF lining, the reduced manpower requirement and increased service life by at least 100% results in substantial total saving for the SLA-92 based castable solution. The work of Duhamel and Verelle was rewarded with the "Young Engineer of the Year Award" of the Institute of Refractories Engineers (IRE) in the year 2000.

Figure 2. Ladle pre-heater refractory lining after 12 month usage without any repair [5]. Lining life achieved was 39 months.

Fundamental properties of a new steel ladle pre-heater technology at CORUS Steelworks IJmuiden are described by de Wit, et al. [5]. The introduction of a new near netshape casting machine required substantial metallurgical process changes and an adjustment of the steel ladle refractory lining. It became necessary to optionally reheat empty steel ladles during the ladle cycles. The focus was on ceramic fiber-free refractory linings of the ladle pre-heater covers. An SLA-92 based castable was selected to cope with the requirements of such linings. In this case a refractoriness of at least 1400°C is required, which is up to 200°C above service temperature, in order to tolerate temperature hot-spots and contribute to long term stability. Additionally low specific weight and low thermal conductivity are important to achieve a low weight of the lining to keep the pre-heater steel construction and the hydraulic power within reasonable limits.

Most important is the resistance to thermal shock to withstand numerous cycles between 1200°C and ambient temperature with rapid, almost immediate, heating up and cooling down. Figure 2 shows the SLA-92 based ladle preheater refractory lining after 12 months use without any repair. In the meantime the linings showed a tremendous service life of up to 3 years without intermediate repairs.

Submerged Entry Nozzle Insulation

The non-fibrous insulation of submerged entry nozzles (SEN) for continuous casting has been investigated by Gotthelf et al. [6]. Positive results have been achieved by coating the nozzles with a SLA-92 based slurry as an alternative to fibrous material wrapping. Figure 3 shows that the spray coating procedure also enables the application of the insulation material in the port-holes of the sub entry nozzle in order to further reduce thermal stress and reduce tendency to failure. The trials showed that SLA-92 based spray coating yields a better thermal insulation in comparison to fibrous material which required doubled thickness (6 mm versus 3 mm) to achieve comparable insulating results.

Figure 3. Submerged nozzle with sprayed on microporous insulation on the surface and in the port holes [6]

Figure 4. Microporous pre-cast shapes for kiln roof, 400x350x160 mm with high alumina anchor brick and two 50 mm calcium-silicate plates on top [7]

Steel Reheating Furnaces

Wuthnow et al. [7] report about the special insulating demands in steel reheating furnaces at Hoesch Hohenlimburg and Thyssen Krupp Stahl AG Bochum. The variety of steel grades produced in those mills require numerous and fast changes of the kiln temperature during use. These steelworks run walking beam furnaces or pusher type furnaces for reheating the steel slabs in temperature ranges from 1080°C to 1350°C. SLA-92 based material was tested in 3 main application areas: Pre-cast shapes for stator and cross beam insulation in walking beam furnace versus refractory ceramic fibers, lightweight gunning mix for bricked furnace roof repairs and pre-cast shapes for furnace roof replacing high alumina insulating bricks. The furnace roofs are lined with lightweight refractories to accelerate the adjustment of kiln temperature between different steel grades to be reheated prior to rolling.

The thermal shock resistance of the refractory lining needs special attention due to frequent process temperature changes and especially during annual or bi-annual maintenance periods with complete cooling down from 1300°C to ambient temperature. SLA-92 based pre-cast shapes for furnace roofs (see Figure 4) were used in comparison to commonly used standard lightweight bricks ASTM class 30, which usually show extensive spalling after 18 months of use. The pre-cast shapes (bulk density at 1.12 g/cm^3) outperform the high alumina insulating bricks (bulk density at 1.08 g/cm^3) with regard to thermal shock resistance and insulating behavior and the SLA-92 based material has been successfully tested over a period of more than 3 years.

A parallel development for stator and cross beam insulation is reported by Kikuchi et al. [8]. The focus was on the replacement of the commonly used fiber insulation in this application. The test results clearly demonstrate the excellent performance of the SLA-92 based insulating castable compared to fiber insulations regarding resistance to slag and spalling. Lining options for stators are pre-cast shapes (see Figure 5) or on site installations. During the introduction period pre-cast shapes have been used. In the meantime the pre-cast shapes are replaced by more cost efficient on-site lining with the additional advantage of an installation without joints. Occasionally superficial cracks have been observed during use.

This has been overcome by replacing calcium aluminate cement with hydratable alumina binder in the insulating castable. Kikuchi et al. report a service life of the lining with SLA-92 of more than 2 years, and it is expected to exceed the service life of the conventional ceramic fiber insulation. The use of SLA-92 in insulating on-site installed castable for stators and cross beams in walking beam furnace is now an established application at a Japanese steel plant.

Figure 5. Pre-cast shape for walking beam reheat furnace stator insulation

Applications in the Petrochemical Industry

Petrochemical applications with direct contact of the refractory lining to the process atmosphere are hydrogen reformers and gasifiers operating with reducing gases containing hydrogen and carbon monoxide gases. Process conditions are pressures around 25 bar and temperatures in the range of 950 – 1100 °C. Additionally, fluid catalytic cracking units ("cat crackers"), ethylene furnaces, pressurized fluid bed boilers, and transfer lines can be included as demanding refractory applications [9].

Important requirements of petrochemical applications are the stability of the refractory oxides against reduction, resistance against CO attack, and abrasion resistance due to high the high velocities of catalyst bearing gas streams in the vessels. Oxides with lower stability like SiO_2 can be reduced by the process gases to gaseous SiO, which afterwards condenses in heat exchangers ("fouling") and reduces the efficiency of this unit. Due to the SiO_2 decomposition the strength of the refractory lining decreases and the porosity increases. The hydrogen attack is discussed in more detail by Tassot et al. [9].

SLA-92 provides an interesting alternative to bubble alumina for high purity insulating materials, which are stable even under severe reducing conditions. Bubble alumina materials have a thermal conductivity of about 1 W/mK, which increases significantly towards higher temperature. SLA-92 materials have a stable low thermal conductivity of about 0.4 W/mK over the entire temperature range.

In a recent investigation by DIFK/Bonn, a SLA-92 based castable was classified for CO resistance to class A after 540°C pre-firing and class B after 1095°C pre-firing according to the norm ASTM C 228.

SLA-92 based materials have been used successfully for petrochemical applications, but this has not been reported in the literature so far.

Applications in the Glass Industry

Windle and Bentley [10] discuss trends in the glass industry for oxy-fuel fired melters, primarily driven by regulation of both NOx and particulate emissions, which show significant reduction utilizing this technology. Besides associated benefits including decrease in fuel, increased melting rate and improved overall productivity, drawbacks are increased alkali concentrations from 1.5 to 6.0 times when compared with conventional melting. This would result in severe wear of traditional silica crowns. Improvements for alkali resistance are achieved by the use of magnesium aluminate spinel for the hot face in crown applications. Regarding the back-up insulation for this spinel lining concept, calcium hexaluminate based insulating bricks were used since they show an improved resistance to alkali attack compared to the previously used calcium-silicate based insulation material. Due to increased temperatures at the outside face of the spinel crown, the high refractoriness of calcium hexaluminate has advantages over commonly used calcium-silicate based insulation material which would run very close to its classification limits [10].

Applications in the Ceramic Industry

Modern ceramic fast firing uses very short cycles, e.g. 60 minutes for ceramic tiles from cold to cold with a maximum temperature of 1140°C, or porcelain 300 minutes from cold to cold with a maximum temperature of 1400°C. Often the kilns are turned off during weekends, which is an additional challenge with regard to the thermal shock resistance of the refractory linings.

In 1999, Stainer and Kremer [11] predicted an interesting potential in ceramic processing using the new microporous calcium hexaluminate SLA-92. The key selection criteria is the excellent thermal shock resistance which is hardly found on other insulating materials at temperatures up to 1450°C.

Periodic kilns used in the ceramic industry benefit from a lowest possible bulk density of the refractory lining because it reduces energy loss through the heat capacity. Pörzgen et al. [12] describe an innovative kiln car lining for decoration firing of ceramic products. The lining is based on an insulating castable with SLA-92 for pre-cast shape manufacturing. The key advantage for this concept is the combination of low thermal conductivity and high thermal shock resistance. This significantly reduces the thermal spalling of the lining compared to conventional insulating bricks but also provides advantages versus refractory ceramic fibers, which de-vitrify during use and become brittle. The increased spalling resistance leads to a reduction in the amount of small particles formed which would otherwise be blown by the burners of the kiln and deposit on the decoration, downgrading the quality of the fired goods and reducing yield.

A new system of kiln car lining concept for decoration firing with SLA-92 pre-cast shapes and refractory plate on top is shown in Figure 6. Firing condition in the kiln is 90 minutes cold to cold with a maximum firing temperature of 1260°C and each kiln car performs 40 cycles per week. The new lining design has been successfully tested over a period for 6 months, namely for about 1000 firing cycles. Excellent thermal properties could be achieved and no damage by thermal shock was observed. Later investigations showed that SLA-92 based material exceeds the lining life of conventional systems, which is between 12-24 months.

Figure 6. (Left) Kiln car lining with microporous pre-cast shapes (white) and refractory plate on top. The bottom layer is made of moler bricks [12]
(Right) Kiln car with decoration firing kiln feed [12]

Overhoff et al.[13] report recently about successful tests with SLA-92 based insulating material in kiln cars for the porcelain industry and special wall bricks for a roller furnace. Particularly for high-alkali, reducing kiln atmospheres and temperatures between 1300 and 1500°C, the material was identified as a highly interesting alternative to lightweight corundum bricks, which have a much higher density and higher thermal conductivity. The excellent thermal shock resistance enabled the production of very complex and large-sized components. Figure 7 shows an application in a special roller kiln using pre-cast shapes as large sized bricks with 250x250x125 mm dimension. Even after 16 months in use, the wall is still in excellent condition.

Figure 7. Wall area of a roller kiln with special pre-cast insulating bricks [13]

Applications in Other Industries

For similar reasons as mentioned above, other industries are discovering innovative uses for SLA-92. Aluminum and other nonferrous metal producers are now testing SLA-92 based refractories for energy savings due to the ability of SLA-92 to withstand harsh reducing environments. In processes that are very sensitive to silica contamination SLA-92 provides a sound alternative to prevent contamination due to this or other impurity.

SLA-92 has a unique feature in that it maintains a very low thermal conductivity even at temperatures near 1500°C (reference Figure 8). The low thermal conductivity from 300°C up to 1400°C is a distinct advantage of SLA-92 based refractory products. This is a result of the high level of microporosity which keeps the dominant heat transfer mechanism (radiation) at increasing temperatures, at low levels. Fiber based products or insulating bricks with comparable refractoriness show an increase in thermal conductivity with rising temperatures. As a result, SLA-92 will provide superior performance in many high-energy intensive processes.

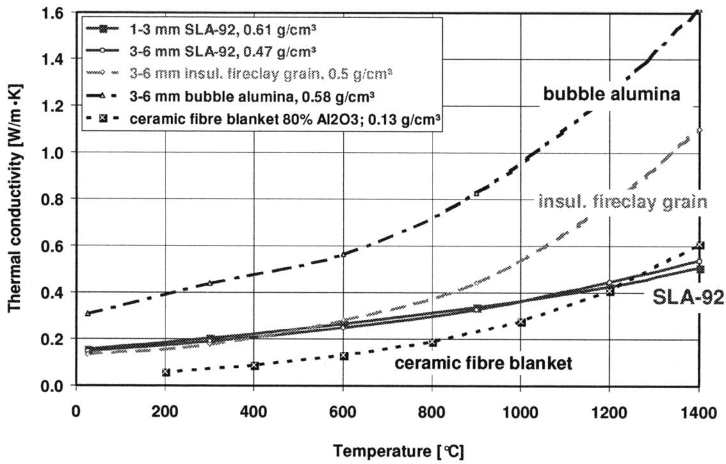

Figure 8. Hot Wire Thermal Conductivity of Select Grains

Conclusion

The comprehensive application experiences with the microporous calcium hexaluminate SLA-92 clearly demonstrate that this insulating aggregate provides both a technical and an economical alternative to conventional refractory ceramic fibers and other insulating materials. As a refractory aggregate it allows the formulation of monolithic concepts for pre-cast shape manufacturing and on-site installations as well as gunning mixes for repairs.

Unmatched properties like high chemical purity, high refractoriness up to 1500 °C, low thermal conductivity at about 0.4 W/mK and high thermal shock resistance are the basis for the outstanding performance observed in various applications over the past years. They led to versatile use in various industries including steel making, glass, ceramics and other refractories and made SLA-92 the material of choice for demanding high temperature insulation. Even under severe conditions, as for example exposure to harsh thermal shock conditions or alkali attack, service life has by far exceeded initial expectations. Obviously the most established and growing application is the insulation lining in steel reheat furnaces for hot rolling mills.

It is expected that applications will further broaden and that SLA-92 will be the basis for other new innovative high temperature insulation concepts for a growing market in all regions world wide.

References

[1] Van Garsel, D.; Swansinger, T.G.; Routschka, G.: New Insulating raw material for High Temperature Applications. 41st International Colloquium on Refractories, Aachen, Sept.1998, p. 122-128.

[2] Van Garsel, D.; Buhr, A.; Gnauck, V.: Long Term High Temperature Stability of Microporous Calcium Hexaluminate Based Insulating Refractories. UNITECR'99 Congress, Berlin, Sept. 1999, proceedings, p. 181-186.

[3] Meyer, F.: BINE Informationsdienst, Neue Wärmedämmstoffe für Hochtemperaturprozesse. Projektinfo 09/04, Fachinformationszentrum Karlsruhe, Büro Bonn, ISSN 0937-8367, p. 1-4.

[4] Duhamel, S.; Verelle, D.: Developments in the Refractory Design of Lids for Ladle Preheating in Dunkirk Steelworks. The Refractories Engineer, Journal of the Institute of Refractories Engineers, March 2000, p. 29-33.

[5] De Wit, T.; Lorenz, W.; Pörzgen, D.; Specht, M.; Buhr, A.: Innovative Ceramic Fiber Free Steel Ladle Preheaters at Corus Steelworks IJmuiden, 44th Colloquium on Refractories, Aachen, Sept.2001, p. 108-112.

[6] Gotthelf, D.; Schrick, G.; Buhr, A.: Faserfreie Isolierung von Tauchausgüssen beim Stranggießen. Stahl und Eisen 121 (2001), No.3, p. 73-78.

[7] Wuthnow, H.; Pötschke, J.; Buhr, A.; Boßelmann, D.; Gerharz, N.; Golder, P.; Grass, J.: Experiences with Microporous Calcium Hexaluminate Insulating Materials in Steel Reheating Furnaces at Hoesch Hohenlimburg and Thyssen Krupp Stahl AG Bochum. 47th Colloquium on Refractories, Aachen, Oct. 2004, p. 198-204.

[8] Kikuchi, T.; Sakamoto, Y.; Fujita, K.: Non-Fibrous Insulating Castable which utilize Micro Porous Aggregate. Fourth International Symposium on Advances in Refractories for the Metallurgical Industries, Hamilton/Canada, Aug. 2004, paper 49.2.

[9] Tassot, P.; Bachmann, E.; Johnson, R.C.: The Influence of Reducing Atmospheres on Monolithic Refractory Linings for Petrochemical Service, UNITECR´01, Cancun, Mexico, Proc. Vol. II, p. 858-871.

[10] Windle, C.J.; Bentley, V.K.: Rebonded Magnesia-Alumina Spinel Products for Oxy-Fuel and Alkali Saturated Atmospheres. UNITECR'99 Congress, Berlin, Sept. 1999, proceedings, p. 219-225.

[11] Stainer, D.; Kremer, R.: Calcium Hexaluminate Products, Development and Application in the High Temperature Industry. UNITECR'99 Congress, Berlin, Sept. 1999, proceedings, p. 187.

[12] Pörzgen, D.; Heide, W.; Buhr, A.: Innovative Refractory Solutions Using a New Microporous Material for Kiln Cars in the Ceramic Industry. CN Ceramic News, Special Refractories, Volume 7, No.2, 2000, p. 68-70.

[13] Overhoff, A.; Buhr, A.; Grass, J.; Wuthnow, H.: New Microporous Materials for Use in Modern Firing Plants. Cfi/Ber. DKG 82 (2005) Nr. 8.

Energy Technology 2010
Edited by: Neale R. Neelameggham, Ramana G. Reddy, Cynthia K. Belt, Ann M. Hagni and Subodh Das
TMS (The Minerals, Metals & Materials Society), 2010

IDENTIFYING SOME POTENTIAL FUTURE HYDROMETALLURGICAL PROCESSES IN TREATMENT OF NICKEL LATERITES

K. Sarveswara Rao
Hydro- & Electro-Metallurgy Department
Institute of Minerals & Materials Technology (CSIR)
Bhubaneswar 751 013, Orissa, India.

Keywords: Nickel laterites, leaching, energy aspects, environmental impacts

Abstract

Existing technologies that currently produce nickel worldwide utilize only about half of the nickel laterite deposits. So, there is a need for a new processing route to recover nickel and cobalt from the remaining laterite ores that are otherwise unutilized. Accordingly, an overview of different existing types of nickel resources and economical processes based on laterite horizons, consumption of energy and reagents and environmental concerns is made in this paper, along with a summary of metallurgical and engineering challenges to be overcome when implementing new technologies and concepts with regard to base metal extraction and sulfur removing efficiency. The paper also compiles and analyzes a range of data concerning energy and greenhouse emissions pertaining to nickel technology.

Introduction

Nickel laterite ores have increasingly become an important source of nickel. As a consequence, many significant achievements regarding laterite project developments are being reported. Nickel can be found within two distinct ore types as sulfide and laterite. Approximately 60% of the world's nickel production originates from sulfide ores. However, laterite ores account for about 70% of the world's nickel reserves and are expected to gain a greater share of total nickel production [1]. Existing technologies that produce nickel worldwide utilize only about half of the nickel laterite deposits. So, there is a need to improve on current laterite processing technologies for recovering nickel and cobalt from the remaining laterite ores that are otherwise unutilized mostly due to complex mineralogy. Accordingly, mining companies are attempting to expand economically viable laterite horizons, lower capital expenditure, lower energy consumption, lower reagent consumption, improve metal recovery and deal with environmental concerns [2].

Materials characterization and the uncertainties associated with conventional leaching kinetics in hydrometallurgy will also play an important role. For example, Prosser [3] has pointed out that the recent approach of researchers has been to obtain valuable evidence for solid reactants and products, by making use a multidisciplinary approach comprising of many techniques. These include optical microscopy; mercury porosimetry; differential thermal analysis and thermogravimetric analysis; X-ray diffraction and spectroscopy;

electrophoresis; various electrochemical techniques including voltammetry, chronopotentiometry and the use of carbon paste electrodes; electron paramagnetic resonance; energy dispersive analysis; low-energy electron microscopy; elastic recoil detection analysis; scanning tunneling microscopy and atomic force microscopy; and isotopes as tracers. Although most of the techniques were originally used by mineralogists, these are now increasingly applied in process mineralogy to more thoroughly assess proposed mechanisms and to minimize uncertainties that might arise in the study of leaching systems.

Similarly, under the experimental conditions used, a multidisciplinary approach ensures elimination of various uncertainties and provides better analysis of kinetic data, in the light of raw material characteristics, experimental variables, solution analysis, nature of products formed, etc., as discussed for lean grade complex sulfide ores [4]. Microscopy, X-ray diffraction (XRD), thermal analysis (TA), surface area measurements, chemical phase analysis and electrochemical measurements were the techniques used for a simultaneous study on characterization of raw materials and leach residues. The same methodology will be very much relevant for characterization of nickel laterites as some TA data are presented later in the text.

In view of abundant literature available on both advantages and disadvantages of various processes, many overviews are published to evaluate the process development target criteria and identify some potential future processes in the treatment of nickel laterites. With the above background, an overview of some recent advances and a comparative evaluation of the process options, and process commercialization issues pertaining to leaching of laterites is made. Energy conservation and environmental compliance are the global aspects/issues and briefly highlighted in the text.

Existing Technologies / Current Practice

Recent publications have reviewed a significant body of literature and examine the most economical commercial processes available to treat nickel laterites [2,5]. The CRU International Report – Nickel Quarterly January 2007 has identified a list of 100 firm, probable, and possible nickel laterite projects through 2015 out of which 54 applied to laterite ores. Those include high-pressure acid leaching (HPAL), 25 (46%); smelting, 17 (31%), heap leaching (HL), 6 (11%) and one each opted for acid chloride leaching (ACLP) and Caron process, respectively. Accordingly, there are three main options: pyrometallurgical (ferro-nickel smelt); Caron (roast-ammonia leaching); and HPAL; fourth route, HL; and fifth option 'others' being under development, such as atmospheric leaching. The trend for new developments clearly points towards HPAL, smelting, and HL technologies. From the process model developed, the costs associated with the ferro-nickel process were observed to be higher in comparison to those associated with HPAL and HL [5].

Similarly, the assessment of technical and economic parameters affecting process selection for treating nickel laterite ores, the selection of the process route to be employed depends on several factors, including the physical and chemical nature of the ore body, weather conditions, and geographic location, which is related to the supply and cost of necessary reagents and energy. Attempts were made to develop a model that summarize the effect of the deposit size and the ore grade on the attractiveness of the high pressure acid leaching (HPAL) and heap leaching (HL) processes for the selected ore. A boundary was shown defining HPAL and HL predominance regions as a function of Ni content and ore body size for the sample. Projects over the boundary line favor the HPAL process and below the line, HL is suitable. The higher the nickel content and the larger the deposit size is, the more favorable the HPAL route becomes. In general, higher nickel prices favor the HPAL route. Smaller deposits and low-grade deposits are better suited for HL, even with increasing nickel prices [5].

Other publication [2] has summarized the new developments as relevant to sulfate based- and chloride based atmospheric leaching processes to treat both limonite and saprolite ores. It also includes a review of the HPAL process and its comparison with the alternative AL processes [6,7]. A combination of HPAL and atmospheric (post-leach) neutralization with saprolite ore, the so-called enhanced pressure acid leach (EPAL) process has also been developed and is currently being applied to the Ravensthorpe project in Western Australia. Here, the limonite and saprolite ore components are treated by HPAL and AL, respectively. A recent patent indicates reaction time of 1-2 h and temperature of $150^{0}C$ to be sufficient to address the concerns of residence time and efficiency of the saprolite neutralization step. It was claimed that the cost of moderate-pressure autoclave systems could be comparable to that of AL reactors and would be much simpler to operate and maintain than the HPAL autoclaves [8].

The authors [2] have also provided a summary of important aspects of the atmospheric chloride leach (ACL) based on the literature: an increased proton activity during leaching due to low water activity as a result of high chloride concentration being used, formation of dehydrated Fe hydrolysis products like hematite under atmospheric pressures, pyrohydrolysis of $MgCl_2$ solution at relatively low temperature of ~$500^{0}C$ to produce highly reactive MgO, and the co-precipitation of water during pyrohydrolysis makes the economics of the process highly sensitive to the major impurities. Another publication [9] elaborates the volatilization of HCl from concentrated chloride solutions to allow the acid to be recycled internally and reduce the need for an expensive pyrohydrolysis step. This technology is said to offer a new alternative in dealing with the leached Fe impurity from laterite ores, but its economical success may hinge heavily on the heat integration efficiency during evaporation to drive Fe(III) hydrolysis, especially since the leaching and liquid/solid separation step may require significantly more dilute environments.

New Developments / Future Outlook

Due to many reasons, there are shortcomings and advantages of the available commercial and pre-commercial processes for treating nickel laterite ores. It is true that no single process developed would be economically superior to handle the full laterite profile at any given geographical location, and hence the need to identify some potential future processes for the recovery of Ni and Co from laterites. In that direction, the authors [2] have described ARNi process, a closed-circuit Mg brine process operating under atmospheric conditions. It operates in the mixed chloride/sulfate brine system and exploits the benefits which both systems offer: a chloride background increases proton activity; suppresses sulfate solubility and produces dehydrated salt; increases the oxidation rate of Fe(II) with oxygen/air; allows for the removal of Fe as hematite at atmospheric pressure; sulfate helps the regeneration of the proton, more so by direct absorption of SO_2 in the solution, and allows for the removal of impurities, such as Na as jarosite. Other ACL process [10] has its own advantages, but has to overcome the limitations of maintaining magnesium concentration of acidic brine and water balance, and treating high magnesium saprolites.

Fundamental studies have demonstrated that hydrochloric acid is more effective than sulfuric acid for the extraction of nickel laterite ores under atmospheric conditions. According to McDonald and Whittington [6,7], current work with nickel laterites and chloride brine media shows promise and development work continues on raw material with low magnesium content. Along with the need to use materials that are corrosion resistant, the maintenance of a proper water balance would seem to be a major issue with all of the atmospheric chloride leaching processes currently under development. The use of organic acid metabolites for the extraction of nickel laterites is still in its infancy. There are also findings in literature that ultrasonic pulses enhanced bio- and chemical processes. More work is needed to examine metal recovery and the effects of recycling of organic acid.

Thermal Analysis (TA) Studies

Earlier, in the Introduction section, it was mentioned that an interdisplinary approach was used for characterization of sulfide raw materials and their leach residues [4]. It also included a TA technique for characterization of raw materials. The same methodology was extended to study the mineralogical characterization of nickel laterites. The TA data [11] thus obtained for three different types of laterite samples (Tisco, OMC and Kansa origin, Orissa, India) are now presented in Table I and discussed.

The TA data of both the bulk and -53 micron TISCO samples at 10^0C/min heating rate suggest two types of dehydration reaction. Those are (a) a broad endothermic depression around 300^0C and a sharper endo- peak at 330-340^{0}C. Thermogravimetric data indicate weight losses of 8.76 and 8.41% for bulk and <53 micron samples, respectively. However, direct heating in a furnace at 950^0C for one hour gives weight loss (LOI) values of 10.48 and 12.44 as shown in Table I.

Table I. Thermal Analysis Data for Different Types of Laterite Samples

	TISCO		OMC		Kansa	
Characteristic features	Bulk	<53 micron	Bulk	<53 micron	Bulk	<53 micron
Temperature range of FeO.OH decomposition (^{0}C)	240-360	245-370	250-355	260-360	230-350	230-345
DTA peak temperature (^{0}C)	335	337	300	310	292	292
TGA weight loss	8.76		3.50		9.30	

(%)		8.41		5.86		10.00
Calculated wt. loss from total Fe_2O_3 (%)	7.35	7.30	3.44	6.02	7.43	7.45
LOI (at 950^0C for one hour) (%)	10.48	12.44	5.95	11.56	12.40	13.34

Differential thermal analysis data of the two samples of OMC show that the decomposition temperature of goethite are lower than those of TISCO samples. The observed loss in weight from TGA agree well with that calculated from the total Fe_2O_3 content (i.e. FeO.OH). However, the LOI value for final fraction shows a much higher result which can be explained due to drastic elimination of chert in the final fraction. Thermal analysis of Kansa sample demonstrate that the DTA peak of goethite decomposition appear at temperature (292^0C) lower than either TISCO or OMC ore. Weight losses observed from TGA traces are as usual higher than those calculated from total Fe_2O_3 but lower than that of LOI data.

The endothermic peak around 300^0C representing goethite-hematite transformation in DTA curve is always lower and complex for poorly crystalline material and with reduction in crystallite size of goethite. For well crystalline goethite the peak is simple and sharp at 385^0C. Kansa samples have peak temperature at 292^0C which points out that the samples contain abundant amorphous or poorly crystalline goethite with low crystallite size. Thermograms of TISCO samples show a humpy peak at 337^0C. These samples contain a larger amount of crystalline goethite with higher crystallite size. XRD data were found useful to supplement the observations made as regards crystallinity. This observation is a typical example showing how TA data are found useful for characterization of laterites having a complex mineralogy.

Environmental Impacts

The environmental impacts involved for pyrometallurgical/hydrometallurgical process of nickel metal production may be expressed as follows.

	GER (MJ/kg)	GMP (kg CO_2e/kg)	AP (kg SO_2e/kg)	SWB (kg/kg)
Flash furnace smelting and Sherrit-Gordon refinery	114	11.4	0.130	65
Pressure acid leaching and SX/EW	194	16.1	-	351

The abbreviated terms are: GER – gross energy, GWP – global warming potential, AP – acidification potential, and SWB – solid waste burden.

The data [12] indicates a much higher SWB for the latter process even after accounting for the lower ore grade. This was attributed to the greater amount of mining waste associated with nickel laterite ores. Hydrometallurgy is found suitable in treatment of wastes and environmental protection because the industry is aware of the fact that recovery and recycling are complementary processes to reduce energy costs. Wastes from a particular step in a process are quite often the inputs for other process steps or industries. Such technologies aim at metal selective leaching and producing easily processable and/or safely stored or disposable residues. The efficiency of such processes has a cost in energy and complexity in view of present growing concerns over global warming, carbon taxes and rising electricity costs. Under the circumstances, there is even more of a focus to limit the energy costs by producing a pure intermediate product that can be sold as a high-value concentrate to a smelter or a hydrometallurgical plant. It appears that more detailed studies in this area should be carried out by researchers in the future.

Interestingly, the Falconbridge nickel smelting technology developed in 2005 was stated to address the key issues of energy efficiency and emissions by using equipment from related processes in other industries already operating under the expected conditions, and all emissions are captured in a single dust-handling system [13]. Higher-value metals, such as nickel and cobalt, when used in products such as nickel-cadmium and lithium-ion batteries, can be valuable sources of feed to nickel smelters. Towards greening of the industry and the role of recycling, a good example is the closure of the sulfur loop in pyrometallurgical operations wherein SO_2 will be captured and made into sulfuric acid, in turn, the acid can be used for leaching of oxide ores or sulfide concentrates [14]. Those smelters are useful conduits for the recycling of PGMs because their downstream refineries are already set up for recovering these elements.

About the context, perspectives and methods of eco-efficiency in primary metals production, van Berkel [15] has indicated that eco-efficiency is aimed at five specific 'resource productivity themes'. Among those, reduction of energy consumption and greenhouse gas emission is identified for reducing process energy requirements, recovery of discharged process heat, and the use of low carbon and renewable fuels and reductants. The QNI yabulu nickel refinery was cited as an example to achieve energy and water synergies through the use of cool boiler feed water to condense still off gases; green water reuse in the nickel thickener circuit and cobalt plant water reuse. It was possible to reduce energy intensity by 2.6%, water use intensity by 9.8% and greenhouse gas intensity by 2.3% for achieving considerable annual savings in operating costs.

Concluding Remarks

The paper has highlighted the shortcomings and advantages of the currently available processes for treating nickel laterite ores. It will be economical to adapt HPAL, EPAL and HL process routes depending on the location, size and classification of nickel resources. In addition, recent trends in the fundamental studies as relevant to chloride based technologies have been described. The environmental impacts related to nickel process technology are also briefly discussed.

References

1. A. Heap, A. Tonks, *"Nickel update, Global Commodities Strategy"*, Citigroup Global Markets 2007.

2. J.D.T. Steyl, M. Pelser, and J.T. Smit, *"Atmospheric leach process for nickel laterite ores"*, Proc. Hydrometallurgy 2008, Sixth International Symposium Honoring Robert S. Shoemaker, Society for Mining, Metallurgy, and Exploration, Inc. (SME), Phoenix, Arizona, August 2008, C.A. Young, P.R. Taylor, C.G. Anderson, and Y. Choi, eds., ISBN 978-0-87335-266-6, p.541.

3. P.A. Prosser, "Review of uncertainty in the collection and interpretation of leaching data", *Hydrometallurgy*, 41(1996):119.

4. K. Sarveswara Rao "The role of solids characterization techniques in the evaluation of ammonia leaching behaviour of complex sulphides", *Min. Proc. Ext. Met. Rev.*, 20 (1999):409.

5. Flavia M.R.S. Xavier and Virginia S.T. Ciminelli, *"Development of a model for the assessment of technical and economic parameters affecting process selection to treat nickel laterite ores"*, Proc. Hydrometallurgy 2008, Sixth International Symposium Honoring Robert S. Shoemaker; C.A. Young, P.R. Taylor, C.G. Anderson and Y. Choi, eds., Society for Mining, Metallurgy, and Exploration, Inc. (SME), ISBN 978-0-87335-266-6, p.532.

6. R.G. McDonald, B.I. Whittington, "Atmospheric acid leaching of nickel laterites review, Part I: Sulphuric acid technologies", *Hydrometallurgy*, 91 (2008):35.

7. R.G. McDonald, B.I. Whittington, "Atmospheric acid leaching of nickel laterites review, Part II: Chloride and bio-technologies", *Hydrometallurgy*, 91 (2008):56.

8. D. Neudorf and D.A. Huggins, *"Method for nickel and cobalt recovery from laterite ores by combination of atmospheric and moderate pressure leaching"*. US Patent 2006/0024224 A1.

9. G.P. Demopoulos, Z. Li, L. Beeze, G. Moldoveanu, T. Cheng and G.B. Harris, *"New technologies for HCl regeneration in chloride hydrometallurgy"*, Proc. EMC 2007, GDMB, Clausthal-Zellerfeld, p.253.

10. B. Harris, C. White, M. Jansen and D. Pursell, *"A new approach to the high concentration chloride leaching of nickel laterites"*. ALTA 2006 Nickel/Cobalt 11. ALTA Metallurgical Services, Melbourne, 20pp.

11. K. Sarveswara Rao et al., *"Nickel resources of sukinda valley Cuttack district, Orissa, India – Their characterization & analysis"*, Project Report No.T/MM/1/JAN./1992, Regional Research Laboratory, Bhubaneswar, 1992.

12. T. Salinovich and P. Stracham, *"Nickel/Cobalt laterites"*, ALTA Metallurgical Services, 1995; a reference from T.E. Norgate, S. Jahanshahi and W.J. Rankin, J Cleaner Production 15 (2007):838.

13. Michael G. King, "Nickel laterite technology – Finally a new dawn", *JOM* (2005 July):35.

14. Michael G. King, "Evolution of technology for extractive metallurgy over the last 50 years- Is the best yet to come", *JOM* (2007 February):21.

15. R. van Berkel, "Eco-efficiency in primary metals production: Context, perspectives and methods", *Resources, Conservation and Recycling*, 51 (2007):511.

Energy Technology 2010:
Conservation, Greenhouse Gas Reduction and Management, Alternative Energy Sources

Alternative Energy Resources for Metals and Materials Production

Organizers

Ann M. Hagni
Neale R. Neelameggham
Aldo Steinfeld
Robert Palumbo

Energy Technology 2010
Edited by: Neale R. Neelameggham, Ramana G. Reddy, Cynthia K. Belt, Ann M. Hagni and Subodh Das
TMS (The Minerals, Metals & Materials Society), 2010

Conceptual Process Plant Utility Schemes for Low Carbon Footprint

Neale R Neelameggham, US Magnesium LLC

Abstract

Most chemical process plants and metallurgical production facilities have some sort of co-generation to minimize process energy costs. Use of Alternative Energy Production from sun and wind power with storage can become an integral part of a process plant utility. This will facilitate reduced fossil fuel usage and help minimize thermal emissions. This will be similar to Pollution Control equipment, which have become an essential part of process plants in the past four decades. Conceptual flow sheets of such Process Utility schemes are presented.

Introduction

Power is supplied to industries, commercial and residential users. Over the past decade, Electric Utility companies have made major inroads, in introducing alternative energy resources in an attempt to somewhat reduce fossil fuel use in power generation. Now, on a much larger scale than in the early 1980s, residential users, motivated by government provided tax credits, have started installing alternative energy generators such as roof top wind mills or roof-top photo-voltaic generators. This will minimize future energy costs, help control thermal emissions etc. This alternative power generation is typically connected to the existing power grid with two way power metering. However, the use of alternative energy generation has not been seen at large industrial sites yet.

Paul Scherer Institute, Switzerland and the NREL laboratories in Colorado, are two of the several institutes which have been developing how solar thermal power can be used to supply the endothermic input energy in chemical and metallurgical processes. The theme of this present symposium addresses the above as well as the energy storage needs for such events to become continuous. This paper emphasizes the concept of how alternate energy can be used in the production of materials on a generic sense wherever there is a need for input energy – beyond endothermic process energy - while addressing how to reduce the thermal and carbon emissions.

The alternative energy generation using wind or solar source is discontinuous due to the nature of the energy source. They are not only discontinuous but also variable. This necessitates storage of generated power for close to continuous use with variable loads. C Smith, et al pointed out in 2006, that 'the electric power industry has taken notice of the expansion of wind and its potential' in a summary article on integrating wind generation into utility systems.[1] In 2009, the 1000 MW wind generation project in Texas had to be postponed to 2013 for lack of transmission.[2] There are only a few large MW solar generation units in the US; the projects being slow mainly from lack of adequate storage and redistribution.

Most often, chemical, metals or materials manufacturing plants have their own partial co-generation of power while using purchased power from power utility companies. The manufacturing companies are motivated by minimizing operating costs by reducing the purchased thermal and electrical energy costs - by using waste heat wherever possible and by generating power using purchased fuel or waste heat if less expensive than purchased power. [3]

In many cases, manufacturing companies have also been able to sell power back to the utility companies using a two-way metering system. But most of these have been based on waste heat re-use co-generation. The section 1302 of the American Recovery and Reinvestment Act of 2009 special tax credits for manufacturing on advanced energy facilities include - technologies that create energy from renewable resources (sun, wind, geothermal and other renewable resources). This paper will describe one of the ways by which a Plant Utility system can be modified with alternative energy generation.

Conventional Process Utility System Boundary

All manufacturing plants have always had the following Utility subsection as part of their architecture: Compressed Air Systems, Water systems, Fuel Systems, Steam Systems and Electric Power Systems. A conceptual schematic of the utility system in a process plant is shown in the Encyclopedia of Chemical Processing and Design. [4] Fig.1 shows a fuel supply input into the Plant utility system boundary.

Fig.**1** **Schematic of common utility systems in Manufacturing** [4]

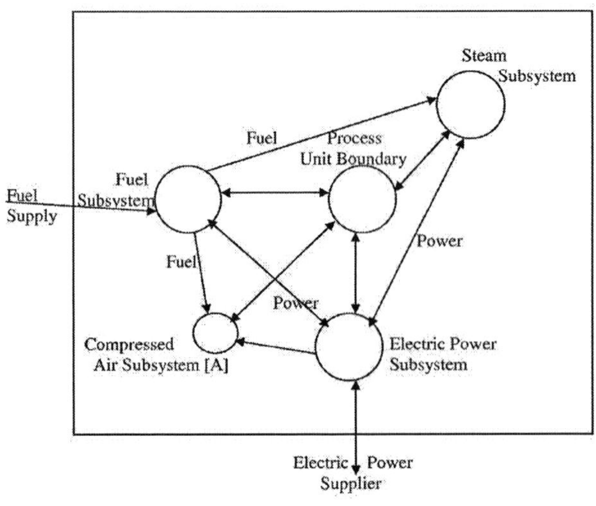

In addition, there is an electric power supplier input-output connection with the Plant utility system boundary. The fuel supply going into the fuel subsystem has 2 outputs - one to the compressed air subsystem and another to steam subsystem. The fuel subsystem also has an input-output connection with the Process unit boundary center and to the electric power subsystem. The steam subsystem has an input - output connection with the process unit boundary and with the power system. The power system has other connections - one output to the compressed air subsystem, one output to the water system [not shown]. In addition it has an output-input connection with the central power unit boundary.

Some Recent Approaches on optimizing Plant Utility System

The awareness of the need to reduce greenhouse gas emissions during the past decade has initiated several studies in Europe and Asia to optimize Plant Utility systems. Tyndall Centre for Climate Change Research in Manchester, England has been one of the forerunners in studying options of using renewable biofuels along with a partial implementation of solar and wind energy input and geological sequestration of CO_2 in over-coming the carbon footprint. Varbonov, et al compared these options with the improvement of energy efficiency of the utility system or the production process. Using economic factors based on 2003 - 2004 on the bio-fuel or alternative wind/ solar generation and 30 to 60 dollars a ton sequestration costs, the authors concluded the following: increasing the system efficiency is the cheapest option for CO_2 abatement - followed by using bio-fuels, CO_2 capture and sequestration. [5] They note, however, that their approach required improvements in the integration of the emissions generation and costing into the overall system model.

Hassan Panjeshahi, et al addressed the sustainability of process industries Utility systems using similar pinch analysis.[6] This study estimated payback periods for six different Renewable energy [RE] sources. Using 2007 costs, the pay back period for the investment costs were found to be 1.5 years for Wind RE to 17.4 years for Tidal RE in the following increasing sequence

Wind < hydropower < geothermal < biomass < solar < tidal.

It is possible that the worldwide increase of commodity prices in early 2008 could have shot the payback period for biomass energy to higher than the 5.4 years shown in this study. The authors point out that by considering CO_2 emission reduction and investment payback simultaneously the order might have changed somewhat as the potential from solar energy's CO_2 emission reduction is quite high compared to biofuels.

Newer Concept of Process Utilities with Renewable Energy Integration

It is suggested that new process plants should attempt to include some element of renewable energy component in their Plant Utility architecture. This can be to the tune of 5 to 20% - which can be motivated by special incentives. These small incremental installations in multiple locations will augment the renewable energy use in a distributed generation basis – similar to what is being done on a residential scale.

Manufacturing industries typically co-generate 5 to 30 per cent of the electric power needed, and are already connected to the grid. It will be simple for both green-field as well as some brown-field conversions to the suggested mode discussed below.

Use of Distributed Energy Storage at Process Industry

There have been several demonstration projects underway where the Utility industry generating renewable power needs to find and install storage devices. Some of the possibilities being perceived are Compressed Air Storage, or pumped water storage [already existing technology].

Residential renewable energy installations do not have the benefit of the economy of scale in utilizing variable energy storage such as compressed air storage and re-use or pumped water storage and re-use, as would an industrial producer who uses larger amounts of electrical or thermal energy.

Many of the mid size manufacturing industries use compressed air for both process air as well as instrument air consuming in the range of 0.5 to 3 MW power. Sufficient storage tanks as buffer between the end use point and compressors are usually present in such systems. A 1 to 6 MW equivalent wind turbine can generate enough mechanical or electrical power to be able to generate the compressed air for the manufacturing process using upgraded storage facilities handling variable generation. Some of the industries which use fuel powered turbo-generators can easily adapt a CAES – adiabatic compressed air energy storage – system, as some of the large Utility plants. [7]

Manufacturing industries can easily be designed to include pumped water storage, or solar thermal storage devices while using renewable energy generators.

Greenhouse Gas Recycle to Fuel At Process Industry

It is also suggested that process industries be designed to include some minimal recycle of greenhouse gases into fuels on-site. This will be similar to the fashion in which the process industries addressed the PM10 pollutants since the 1970s by installing scrubbers. These can be considered as GHG Emissions control devices.

Fundamental concepts of soda-fuel metallurgy have noted this possibility by use of electrolytic action, which requires energy - the source of which can be hybrid wind-solar renewable energy generators.[8] Solar thermal recycle of flue gas by using biomass is another example. [9] Here the term soda refers to the water vapor – carbon dioxide mixture in a flue gas [soda as in carbonated water]. The storage of such variable renewable generating units will be the soda-fuel [H_2O-CO_2] conversion from all or some of the flue gases.
Many in Process Industries are aware of the benefits of regeneration of reagents in cutting down the purchase of reagents to a small fraction. Solvent extraction stripping cycle regeneration using lower priced acid or alkali reduces the expensive organic reagent's make up to 0.5 to 2% of stoichiometric requirement. Titanium metal production by Kroll process where the magnesium reductant is recycled by electrolysis making the makeup requirements to 1 to 5% of stoichiometric. Similar approaches should help reduce input fossil fuel use to a small fraction – which in turn will cause the carbon dioxide and water vapor green-house gas emissions to a small fraction. Indirectly, renewable energy gets used in the process which requires the use of thermal energy from conventional hydro-carbon fuels.

Many industries have also found out it is less expensive to convert the processes to release concentrated effluent, so that capturing and turning them into a useful product will be less expensive. Typical example of this is what the copper industry had accomplished by converting from the air operated converters of the 1950's to oxygen operated conversion. This facilitated capture of more SO2 from being released into the atmosphere as sulfuric acid. This also obviated the need for eliminating or minimizing scrubbing using costly alkaline reagents which resulted only in wasteful aqueous effluents instead of a marketable by-product.

The Renewable Energy - low carbon foot print Process Utility Boundary concept suggested above is shown in Fig.2

Fig.2. Low carbon footprint Process Utility System Concept

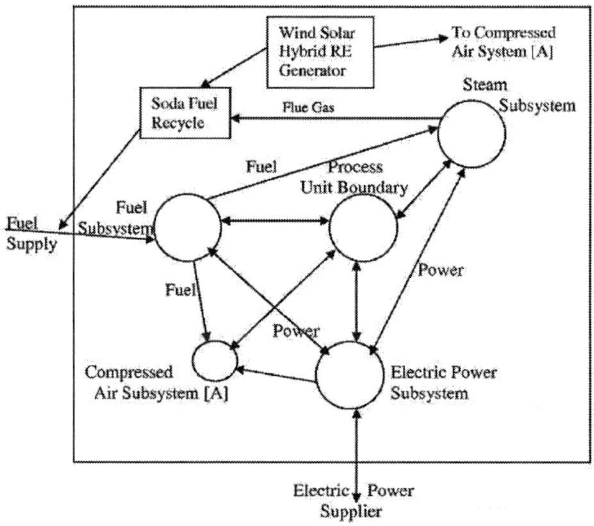

The source reduction from flue gases by introducing useful recycle will have greater benefits than having to release flue gases into the atmosphere and have them take a longer recycle path through- the so-called carbon neutral bio-fuels. In some cases this can also lead to marketing of fuel similar to power re-sale by the consuming industries. In addition this would alleviate the need for CO_2 sequestration such as geological storage which can consume compression and transportation energy of about 1 ton equivalent of CO_2 emissions for every 2 tons sequestered. [10] Instead of such wasteful consumption of energy for sequestration, we establish a method of minimizing fossil fuel use - by on-site application of renewable energy resources.

Reduction of Greenhouse Gas Emissions; Transmission Line Loss Equivalent

When the manufacturing industry installs a renewable power generation on-site, the reduction in the transmission load from Utility industries - is another carbon foot print lowering technique.

This is also true of renewable energy installations at thousands of residential and small commercial facilities.

If we have superconductors used in transmission we may not have this resistive loss... It should be noted that the transmission load reduction is in cutting down the greenhouse gas generation equivalent of the wasted energy in making up for the resistance heat losses in the several miles [in some cases several hundreds of miles – California consumes power generated as far away from Colorado or Washington States] of conductor between the Utility Industry and the consuming manufacturing industry. Manufacturing industry is usually charged this power loss in their tariff. The process optimization models should include this cost and the avoided greenhouse gas emission costs by avoiding the resistive losses. This can be done through carbon credits to the manufacturing industry.

Conclusion

Implementation of such fractional Renewable energy Generation at manufacturing industries will be able to reap faster benefits in the green-house gas reduction approaches - beyond what is being done on the Residential and small commercial levels with grid connection.
The reason for faster benefits is obvious from the distributed nature of all the material production plants world-wide.

References

1. http://ww.uwig.org/WindIntegration-NAW92006.pdf.
2. http://headlines.ocregister.com/articles/pickens-33083-project-wind.html
3. Roger Swenson, R.Neelameggham, ' Magnesium Producer Relies on Distributed Generation with Combined Heat and Power, Distributed Energy' - Combined Heat and Power- A supplement to Energy Matters, Department of Energy publications, Fall 2001
4. Ravinath, Dale J. Libby and Howard J. Dubon, 'Optimization Techniques, Joint Process Units and Utility System', Encyclopedia of Chemical Processing and Design, Vol.32, ed. John J Mcketta Jr., Marcel Dekker, 1990, p.465-470.
5. Peter Varbanov, Simon Perry, Jiri Klemes, Robin Smith, 'Synthesis of industrial utility systems; cost-effective decarbonization', Applied Thermal Engineering Sept. 2004.
6. Hassan Panjeshahi, Lena Ahmadi and Simon Perry, Proceedings of the World Congress on Engineering and Computer Science, 2008 - October 22-24, 2008, San Francisco.
7. William F. Pickard, Nicholas J. Hansing, and Amy Q. Sher, 'Can large-scale advanced-adiabatic compressed air energy storage be justified economically in an age of sustainable energy?', Journal of Renewable and Sustainable Energy, 1. 033102 (2009)
8. Neelameggham, 'Soda-Fuel Cycle Metallurgy - Choices for CO_2 Reduction' , CO_2 Reduction Metallurgy- Symposium Proceedings, ed. Neale R. Neelameggham and Ramana G Reddy, 2008 TMS Annual Meeting, New Orleans, p. 135-146.
9. Neelameggham, ' Solar Pyrometallurgical Recycling of CO2 - H2O [soda Reactants]', REWAS 2008, ed. B. Mishra, Ludwig and S.K.Das, Cancun, p. 1815-1820.
10. Milind Deo, Private Communication - Dept. of Chem. Eng. University of Utah - 2007

Energy Technology 2010
Edited by: Neale R. Neelameggham, Ramana G. Reddy, Cynthia K. Belt, Ann M. Hagni and Subodh Das
TMS (The Minerals, Metals & Materials Society), 2010

METAL FERRITE SPINELS FOR SOLAR-THERMAL WATER SPLITTING REDOX CYCLES

Jonathan R. Scheffe, Melinda Channel, Alan W. Weimer

Department of Chemical and Biological Engineering
University of Colorado
Boulder, CO 80309-0424 (U.S.A.)

Keywords: ferrite, hydrogen, solar, water splitting

Abstract

There is a need to develop cost effective renewable energy processes for the production of clean fuels that can simultaneously mitigate the U.S. energy/global environmental security issue. If H_2 can be obtained cost effectively via the splitting of water using concentrated sunlight, then it's possible to operate a fuel cell using renewable H_2 or to synthesize other fuels based on renewable H_2 for generating electricity or for running internal combustion engines. For example, H_2 can be reacted with sequestered carbon dioxide (CO_2) to produce syngas (H_2/CO) via reverse water gas shift (RWGS), i.e. $H_2 + CO_2 \rightarrow CO + H_2O$. The syngas can be catalytically methanated to form synthetic natural gas (SNG) or it can be catalytically reformed to produce liquid fuels such as "green" gasoline via the MTG process (syngas→ methanol → gasoline). The desert SW United States is home to world class sunlight. Fundamental ferrite thermodynamics will be reviewed, an example of the beneficial effect of ferrite surface area on reaction kinetics will shown, and process economics will be investigated for a 100,000 kg/day facility to produce H_2.

Introduction and Background Thermodynamics

Spinel ferrites of the form $M_xFe_{3-x}O_4$, where M generally represents Ni, Zn, Co, Mn, or other transition metals, have been shown to be capable of splitting water according to the two step cycle shown below:

$$M_xFe_{3-x}O_4 + \text{solar thermal energy} \rightarrow xMO + (3-x)FeO + 0.5\ O_2 \quad \textbf{(1)}$$

$$xMO + (3-x)FeO + H_2O \rightarrow M_xFe_{3-x}O_4 + 2H_2 \quad \textbf{(2)}$$

This is an inherently clean and sustainable process, as the only net inputs are solar energy and water, and the net outputs are hydrogen and oxygen. Ferrite water splitting cycles are advantages compared to other high temperature water splitting cycles, such as the ZnO/Zn and Mn_2O_3/MnO cycles, because thermal reduction has been demonstrated at significantly lower temperatures[1, 2]. As a result, overall efficiency may be increased due to the fact that radiation losses increase as the fourth power of temperature, and material considerations for reactor design should be more flexible. Additionally, cycle repeatability has been demonstrated with relative ease as compared to other thermochemical cycles [3-6]. However, the amount of hydrogen per mole that is generated is significantly less than for other high temperature thermochemical cycles due to the fact that only Fe^{3+} is capable of being reduced at the temperatures of interest[6]. In addition, water oxidation kinetics appear to be very slow, likely due to diffusion or surface area limitations created by near surface oxidation [6, 7].

Currently, little is understood about the factors that control the reaction rates and conversion efficiencies of ferrite water splitting cycles. To date, most literature has dealt with the effects of the substituted metals in $M_xFe_{3-x}O_4$ and their stoichiometry [3, 4], synthesis methods[4, 8, 9], and various substrates[3, 6] on the total amount of H_2 produced and their ability to be cycled a number of times. The basic thermodynamics of the ferrite cycle is illustrated in Figure 1. It is

Figure 1. Fe_3O_4 Redox Thermodynamics

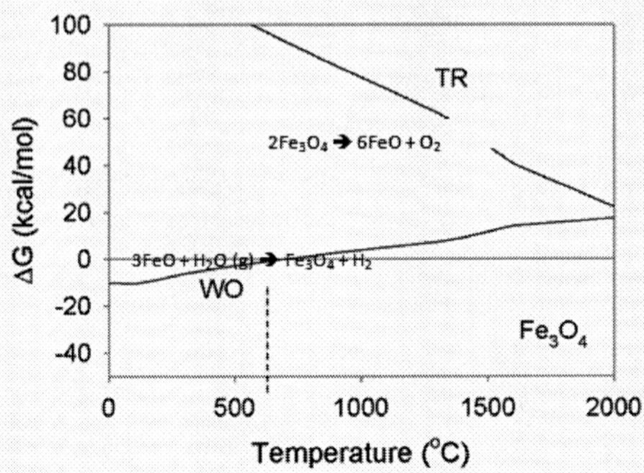

Figure 2. Co_3O_4 Redox Thermodynamics

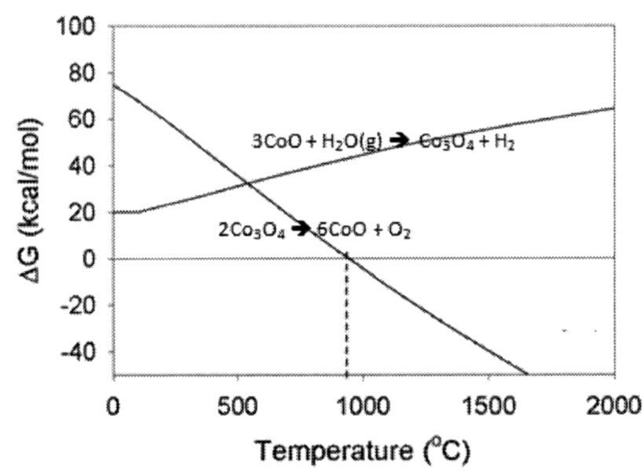

clear from the figure that thermal reduction of Fe_3O_4 is not favorable at desired temperatures below the melting temperature of wustite (FeO). However, the water oxidation step is thermodynamically favorable at relatively low temperature. A similar representation of the thermodynamics for the CoO cycle water splitting cycle is shown in Figure 2. Here, the thermodynamics of the thermal reduction is favorable at desirable temperatures, but, the corresponding oxidation step is not. This behavior has inspired researchers to investigate mixed metal oxides for water splitting. The free energy change for the mixed oxide $CoFe_2O_4$ is shown in Figure 3. It is clear that the oxidation step is favorable, but that the thermal reduction cannot proceed to completion. In fact, CoO is preferred thermodynamically over FeO.

Allendorf *et. al.* [10] recently reported on the thermodynamic effects of Fe, Co, Ni, and Zn

Figure 3. $CoFe_2O_4$ Redox Thermodynamics

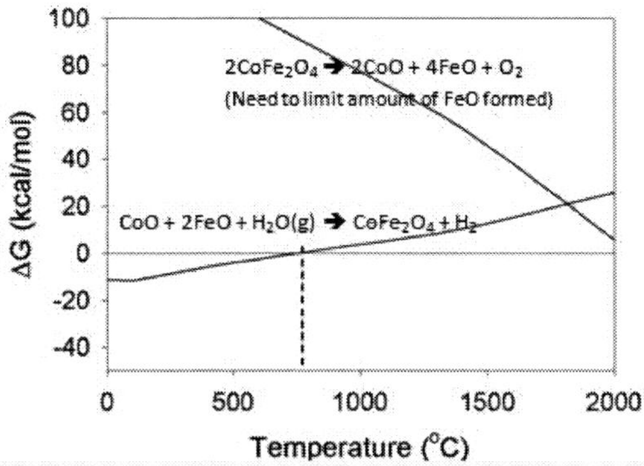

Figure 4. Approach - Free Energy Minimization Theoretical Limit (P = 0.001 MPa)

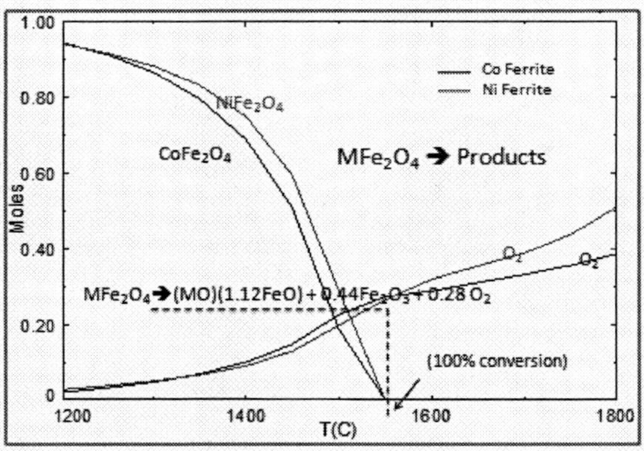

substitution in $M_xFe_{3-x}O_4$, and observed that x and M greatly influence the equilibrium of this reaction for a given reduction temperature [10]. Co and Ni ferrites were both reduced to a greater extent at temperatures of interest (~1400 °C), and it was shown that Fe_3O_4 does not fully reduce until after the melting temperature of wustite (FeO). A Gibbs free energy minimization calculation (performed using FactSage in an analogous manner to Allendorf *et. al.* [10]) shows that both $CoFe_2O_4$ and $NiFe_2O_4$ reduce at lower temperatures than Fe_3O_4, as Allendorf *et. al.* [10] have shown. However, what is interesting to note is the fact that when the ferrite is fully reduced (moles spinel = 0), only ~0.28 moles of O_2 exist at equilibrium (Figure 4), rather than the expected 0.5 shown in Equation 1. This is due to the fact that a significant amount of Fe_2O_3 is expected to exist at equilibrium, limiting the amount of O_2 that can be evolved. Kodama *et. al.* have observed greater cyclical stability and more H_2 production when depositing Ni, Co, and Fe ferrites on ZrO_2 and YSZ supports [3-5]. It was observed that Fe^{2+} forms a solid solution with the substrate during thermal reduction and can be subsequently oxidized to reform the ferrite. Additionally, x and M in $M_xFe_{3-x}O_4$ affected the reported conversions.

Synthesis and Kinetics Considerations

Though thermodynamics and the aforementioned factors have been studied, the kinetics and mechanisms of these reactions have yet to be explored in detail. Rapid oxidation kinetics are desired in order to facilitate high throughput of H_2, as it has been observed that the decomposition reaction is considerably faster than the water oxidation [6, 7]. Charvin *et. al.* observed a strong dependence of Fe_3O_4 particle size on the H_2 reaction rate during water oxidation experiments [11]. As the particle size was increased, slower rates were observed. It was hypothesized that this was due to an oxide barrier formed on the surface of the particles, hindering further oxidation of the bulk. Similar behavior has been observed for CO_2 splitting on chemically reduced $CoFe_2O_4$ nanoparticles [12]. Though, it was hypothesized that this phenomena was due to bulk diffusion limitations, there was no evidence that it was not more dependent on surface area. Ultimately, the economic viability of this thermochemical cycle is dependent on relatively fast kinetics, because, if cycle times can be decreased considerably, it is possible to run many more cycles during limited sunlight hours to produce a given amount of H_2 and requiring less ferrite material (a capital cost).

Various methods have been used to synthesize ferrites for these cycles, including solid state synthesis (SSS) [8, 13-15], co-precipitation[6], aerial oxidation method[4], aerosol spray pyrolysis[8], and self-propagating high-temperature synthesis[8]. SSS is perhaps the simplest of these techniques, but it can be difficult to control particle sizes due to high temperature sintering that occurs during synthesis. This can result in decreased activity of the ferrites, possibly due to the fact that specific surface area plays an important role in either the reduction or oxidation steps [3-5]. Both coprecipitation and aerial oxidation have been used to successfully deposit ferrites on supports, such as ZrO_2 and YSZ, and were shown to increase the reactivity of the ferrites over several cycles [3, 6]. This was attributed to a reduction in high temperature sintering. In this paper, films of $CoFe_2O_4$ are synthesized by first depositing Fe_2O_3 and CoO on ZrO_2 nanoparticles in alternating layers by atomic layer deposition (ALD) in a manner similar to that discussed by Scheffe et al. for Fe_2O_3[16]. The coated nanoparticles are then formed into disc shaped substrates using poly(methyl methacrylate), i.e. PMMA, as a binder. The substrates were heat treated at 700°C in a ZrO_2 crucible to "burn out" the PMMA, leaving a porous structure. Calcination at 1100°C resulted in the formation of $CoFe_2O_4$ which was verified by X-ray diffraction (XRD).

These high surface area ALD prepared materials are compared to standard co-precipitated materials in their ability to split water. In comparing samples, the ALD synthesized samples have a much larger surface area (2.62 m^2/g vs. <0.5m^2/g) and nano-scale thicknesses compared to the co-precipitated samples. The peak H_2 reaction rate of the ALD samples (2.2 wt% loading on ZrO2) is 9.5 μmoles H_2/s/g, whereas the peak rate of the co-precipitated (25 wt% loading on ZrO2) sample was only 1.6 μmoles H_2/s/g, as seen in Figure 5. Additionally, the reaction appears

Figure 5. Comparison of Observed Oxidation Reaction Rates

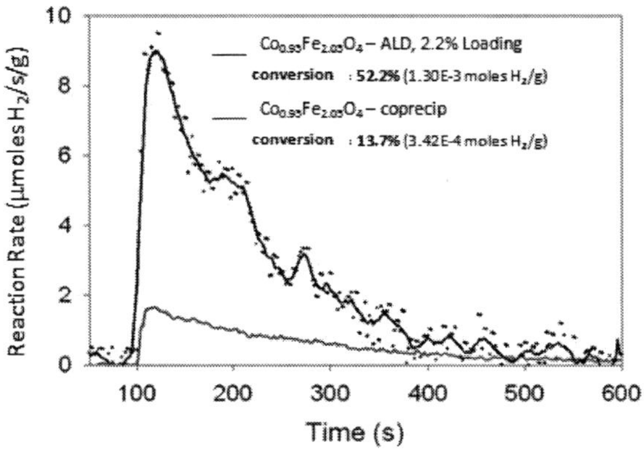

to go to completion after 450 s. After the water is diverted from the reactor after 1600 s (not shown) there was no decrease in the H_2 signal as there was with the co-precipitated sample. The total amount of hydrogen produced using the ALD sample was 1.30E-3 moles H_2/g ferrite, whereas 3.42E-4 moles H_2/g ferrite was produced using the co-precipitated sample. The maximum theoretical amount of H_2 produced from $CoFe_2O_4$, assuming .28 moles O_2/mole ferrite were evolved (see Figure 4) and then all of the Fe^{2+} was re-oxidized during the WO step, is 2.56 E-3. Therefore, respective conversions of each sample are 13.3% for the co-precipitated sample and 50.1% for the ALD prepared sample.

Preliminary Process Economics

A preliminary economic evaluation was carried out using the DOE H2A process economics program to estimate the selling price of plant gate hydrogen (H_2) at 300psig for a 100,000 kg/day solar-thermal facility to split water. A conventional bulk cobalt ferrite spinel material ($NiFe_2O_4$) is considered for the process economics evaluation. The process considered is one in which ferrite pellets flow downward through multiple silicon carbide (SiC) reaction tubes (moving bed reactor) in a multi-tubular solar-thermal reactor and are reduced, releasing O_2, while being heated to 1400°C. The O_2 flows countercurrently upward and is removed with a vacuum pump. The solids are held in an "adiabatic" storage/feed tank at 1000°C and then oxidized with steam to produce H_2 and to regenerate the ferrite spinel material. The envisioned process is shown schematically in Figure 6. The steam oxidation is exothermic, providing heat, and the regenerated spinel is held in another "adiabatic" storage tank at 1000°C until fed back into the solar-thermal reactor. There is no heat recovery assumed between the 1400°C and 1000°C Redox steps in the cycle. It is assumed that eight (8) Redox cycles are carried out each day, 365 days/yr, but that the available sunlight is from actual historical data in Daggett, CA (Mohave Desert). This solar design strategy is consistent with that used at NREL and Sandia. An AspenPLUS™ simulation was carried out for the process.

Figure 6. Base Case Process Flow Diagram

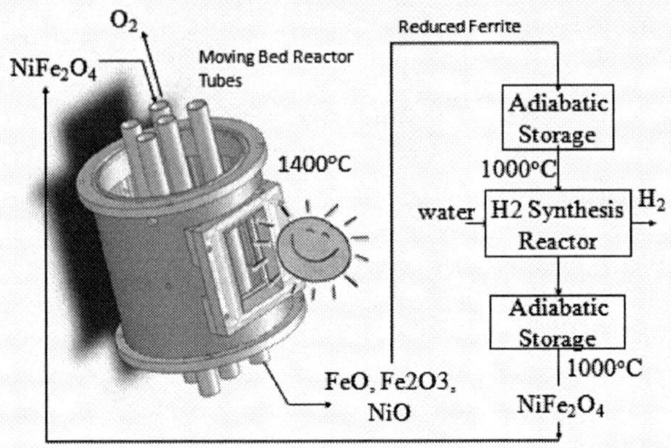

Table 1. Annual Energy Requirements

$$NiFe_2O_4 \rightarrow 1.2\ FeO + 0.4\ Fe_2O_3 + NiO + 0.3\ O_2$$

$\Delta H_{rxn} = 214.6$ kJ/mol (FactSage™ results at 0.1 MPa & 1400°C)

Solar Reactor (GWhr/yr)	Reduction Conversion		
	35%	70%	100%
Overall Heat of Reaction	1,799	1,799	1,799
Sensible Heat	2,044	1,022	715
Total Energy Required	3,843	2,821	2,514

(Cavity operates at 3,868 suns, losses are primarily radiative)

A summary of the energy balance for the process is shown in Table 1 for 35%, 70% and 100% conversion of the reduction reaction (the oxidation step is assumed to be 100% for each case). The base case is taken as 70% conversion. The heat of reduction using ASPENPlus™/FACTSage™ is 214.6 kJ/mol for atmospheric pressure and 1400°C processing. The cavity operates at 3868 suns (i.e. kW/m^2). The overall annual energy requirement to produce 100,000 kg H_2/day is 1,799 GWhr/yr. The sensible heat required to heat the ferrite from 1000°C to 1400°C for reaction is 1,022 GWhr/yr if 70% conversion is achieved in the reduction reaction. The sensible heat requirement increases to 2,044 GWhr/yr if we can achieve only 35% conversion per pass (i.e. more ferrite must be processed to provide the 100,000 kg/day of H_2). If on the other hand, we can achieve 100% reduction of the ferrite, then the sensible heat requirement decreases to 715 GWhr/yr. The total energy required for the base case 70% conversion is 2,821 GWhr/yr. These results are summarized in Table 1.

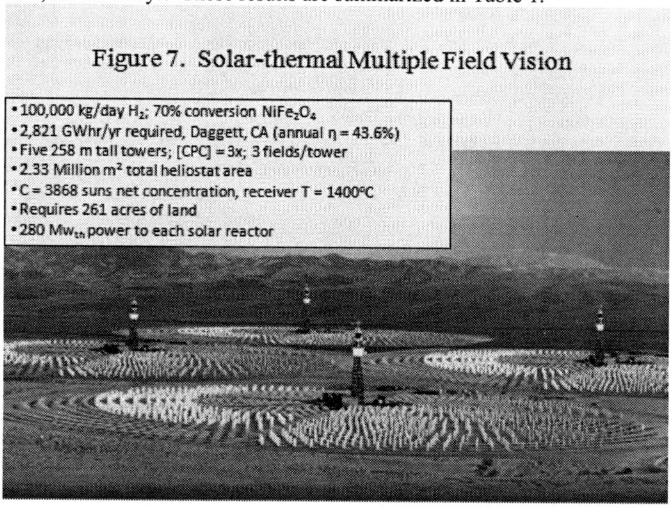

Figure 7. Solar-thermal Multiple Field Vision

Table 2. Cycle Efficiencies

		Reduction Conversion		
		35%	70%	100%
Base Case	LHV	31.7%	43.1%	48.4%
	HHV	37.4%	50.9%	57.1%
By-product Case	LHV	35.8%	45.8%	50.2%
	HHV	41.5%	53.5%	58.9%

A vision for the overall process implementation is shown in Figure 7 where multiple solar fields are shown for the plant. The plant requires five (5) 258 m tall towers and associated heliostat fields. Each tower is equipped with a multi-tubular solar reactor system operating at 1400°C. Concentrated sunlight is provided at 3868 suns net concentration using secondary concentrators on the receivers. The overall annual efficiency for the sun to the receiver is 43.6%. The process requires 2.33 million m^2 of heliostat to provide the 2,821 GWhr/yr of energy. The total land space requirement is 261 acres and each tower is providing 280 $kW_{thermal}$ power to each receiver.

The cycle thermal efficiencies are calculated using the AspenPLUS™ simulation results and are summarized in Table 2. For the base case of 70% conversion and assuming no heat recovery between Redox steps and no byproduct electricity, the lower heating value based process thermal efficiency is 43.1%. This increases to 45.8% (LHV) if electricity is made and sold from heat generated via the exothermic oxidation step at 1000°C. The overall efficiency for sunlight to H_2 for the process is the multiple of the solar efficiency (i.e. 43.6% annual average) and the thermal efficiency (43.1%), or 18.8% based on LHV. If we consider supplying electricity, the overall process efficiency increases to 20%. These overall efficiencies are consistent with efficiencies realized for concentrated solar power (CSP) facilities in Seville, Spain and Solar II in Daggett, CA. The thermal efficiencies for various scenarios are summarized in Table 2.

Process economics were carried out. The major H2A cost assumptions are summarized in Table 3. A key consideration is the DOE cost targets for installed heliostat fields. Prices used are \$180/m^2 today with the targeted plant gate H_2 selling price of \$10/kg H_2. These costs become more stringent with a target of \$140/m^2 for heliostats installed and a selling price of H_2 at \$6/kg in 2012. The 2017 target is \$94/m^2 with a plant gate H_2 selling price of \$3/kg.

Table 3. Major Cost Assumptions

• 3 year construction • 40 year plant life • 20 year MACRS depreciation • 1.9% inflation (H2A) • 10% after tax real IRR (H2A) • 38.9% total tax (H2A) • Engineering & Design = 17.7% of direct capital • Process Contingency = 16.6% of direct capital • No carbon avoidance credits Byproduct Case Only • O_2 sold for \$0.02/kg (H2A) • Electricity sold for \$0.07/kWhr (H2A)	• 2008 Case – \$180/m^2 heliostat cost – \$10/kg H_2 • 2012 Case – \$140/m^2 heliostat cost – \$6/kg H_2 • 2017 Case – \$94/m^2 heliostat cost – \$3/kg H_2 • H_2 delivered at 300 psig

Various economic scenarios were carried out. One economic analysis was done to determine the allowable cost of ferrite to hit a selling price of \$4/kg H_2 for the 2017 byproduct case for selling electricity and O_2. Reduction conversion of ferrite was 70%. The ferrite is a capital cost to the plant since it is not used during the reaction, but rather serves as a catalyst. The capital cost of the 100,000 kg/day H_2 plant is \$654M with ferrite costs comprising 37% of the total allowable capital to hit the \$4/kg H_2 price target with a 10% IRR. The allowable ferrite cost is \$69/kg. The actual cost of NiO and Fe_2O_3 in the quantities to produce the ferrite is \$8.67/kg ferrite, hence, making the process appear very feasible. It is important to note that the solar components –heliostats, towers, receivers comprise 55% of the total capital. On the revenue side, the sales of H_2 represent 93% of the revenue to the plant. The remaining 10% is electricity (3%) and O_2 (4%). A number of other economic scenarios are presented in Table 4 using the cost of ferrite as \$8.67/kg. The price of H_2 for a 10% IRR from H2A is as low as \$2.19/kg H_2 if 100% reduction

Table 4. H_2 Selling Price for 10% IRR ($/kg)

		Ferrite Cost is $8.67/kg ferrite		
		35% Reduction Conversion	70% Reduction Conversion	100% Reduction Conversion
Base Case	2008	$6.52	$4.44	$3.79
	2012	$5.68	$3.82	$3.24
	2017	$4.43	$2.90	$2.42
By-product Case	2008	$6.09	$4.16	$3.56
	2012	$5.25	$3.54	$3.01
	2017	$3.98	$2.61	$2.19

can be achieved for the 2017 byproduct case. For the base case with no byproducts and 70% conversion, the H2A economics hit the DOE price targets for all years with a selling price of $2.90/kg in 2017.

These economics results are very encouraging and indicate that if ferrite can be produced for $8.67/kg and if 70% conversion can be achieved in the reduction reaction, that the process is both technically and economically feasible and will exceed the DOE cost target of $3/kg H_2 in 2017.

References

[1] Perkins, C., and A.W. Weimer," Likely near-term solar-thermal water splitting technologies", *International Journal Of Hydrogen Energy* Vol. 29, No. 15, 2004, pp. 1587-1599.

[2] Sturzenegger, M., and P. Nuesch," Efficiency analysis for a manganese-oxide-based thermochemical cycle", *Energy* Vol. 24, No. 11, 1999, pp. 959-970.

[3] Kodama, T., N. Gokon, and R. Yamamoto," Thermochemical two-step water splitting by ZrO2-supported NixFe3-xO4 for solar hydrogen production", *Solar Energy* Vol. 82, No. 1, 2008, pp. 73-79.

[4] Kodama, T., Y. Kondoh, R. Yamamoto, H. Andou, and N. Satou," Thermochemical hydrogen production by a redox system of ZrO2-supported Co(II)-ferrite", *Solar Energy* Vol. 78, No. 5, 2005, pp. 623-631.

[5] Kodama, T., Y. Nakamuro, and T. Mizuno," A two-step thermochemical water splitting by iron-oxide on stabilized zirconia", *Journal of Solar Energy Engineering-Transactions of the Asme* Vol. 128, No. 1, 2006, pp. 3-7.

[6] Miller, J.E., M.D. Allendorf, R.B. Diver, L.R. Evans, N.P. Siegel, and J.N. Stuecker," Metal oxide composites and structures for ultra-high temperature solar thermochemical cycles", *Journal of Materials Science* Vol. 43, No. 14, 2008, pp. 4714-4728.

[7] Gokon, N., T. Hasegawa, S. Takahashi, and T. Kodama," Thermochemical two-step water-splitting for hydrogen production using Fe-YSZ particles and a ceramic foam device", *Energy* Vol. 33, No. 9, 2008, pp. 1407-1416.
[8] Agrafiotis, C., M. Roeb, A.G. Konstandopoulos, L. Nalbandian, V.T. Zaspalis, C. Sattler, P. Stobbe, and A.M. Steele," Solar water splitting for hydrogen production with monolithic reactors", *Solar Energy* Vol. 79, No. 4, 2005, pp. 409-421.
[9] Lorentzou, S., C.C. Agrafiotis, and A.G. Konstandopoulos," Aerosol spray pyrolysis synthesis of water-splitting ferrites for solar hydrogen production", *Granular Matter* Vol. 10, No. 2, 2008, pp. 113-122.
[10] Allendorf, M.D.," Two-Step Water Splitting Using Mixed-Metal Ferrites: Thermodynamic Analysis and Characterization of Synthesized Materials", *Energy & Fuels* Vol. 22, No. 6, 2008, pp. 4115-4124.
[11] Charvin, P., S. Abanades, G. Flamant, and F. Lemort," Two-step water splitting thermochemical cycle based on iron oxide redox pair for solar hydrogen production", *Energy* Vol. 32, No. 7, 2007, pp. 1124-1133.
[12] Khedr, M.H., A.A. Omar, and S.A. Abdel-Moaty," Reduction of carbon dioxide into carbon by freshly reduced CoFe2O4 nanoparticles", *Materials Science and Engineering a-Structural Materials Properties Microstructure and Processing* Vol. 432, No. 1-2, 2006, pp. 26-33.
[13] Tamaura, Y., A. Steinfeld, P. Kuhn, and K. Ehrensberger," Production Of Solar Hydrogen By A Novel, 2-Step, Water-Splitting Thermochemical Cycle", *Energy* Vol. 20, No. 4, 1995, pp. 325-330.
[14] Tamaura, Y., Y. Ueda, J. Matsunami, N. Hasegawa, M. Nezuka, T. Sano, and M. Tsuji, "Solar hydrogen production by using ferrites", 1999, pp. 55-57.
[15] Han, S.B., T.B. Kang, O.S. Joo, and K.D. Jung," Water splitting for hydrogen production with ferrites", *Solar Energy* Vol. 81, No. 5, 2007, pp. 623-628.
[16] Scheffe, J.R., A. Frances, D.M. King, X.H. Liang, B.A. Branch, A.S. Cavanagh, S.M. George, and A.W. Weimer," Atomic layer deposition of iron(III) oxide on zirconia nanoparticles in a fluidized bed reactor using ferrocene and oxygen", *Thin Solid Films* Vol. 517, No. 6, 2009, pp. 1874-1879.

Energy Technology 2010
Edited by: Neale R. Neelameggham, Ramana G. Reddy, Cynthia K. Belt, Ann M. Hagni and Subodh Das
TMS (The Minerals, Metals & Materials Society), 2010

APPLICATIONS OF CONCENTRATING SOLAR POWER IN MATERIALS PRODUCTION

Daniel P. Cook, Joseph Kopp, Jordan Mayorga, Robert Boehm

Department of Mechanical Engineering, University of Nevada, Las Vegas, 89154, USA

Keywords: Concentrating Solar Power, Nevada Solar-1

Abstract

Solar generated power has long been considered a useful energy source only for low-energy density and non-mission critical use, such as providing electricity for residential applications. The success of Concentrating Solar Power (CSP) facilities, such as the 64 MW Nevada Solar-1 near Las Vegas, have demonstrated that solar does have application to high energy density industries such as materials and chemical production. This paper will discuss the current operating configuration of the Solar-1 facility and show how proposed modifications, such as addition of thermal storage and use of liquid salts as the working fluid, can further improve the efficiency and energy production of the plant.

Introduction

The United States as a nation is the largest consumer of energy in the world at an estimated 29 trillion kWh per year[1], which is roughly 15% of total global energy consumption[2]. On a per capita basis, the United States also leads most of the world, using an average of 96,000 kWh per person[3]. Roughly 40% of the energy consumed in the United States was produced from petroleum, 23% from coal, 23% from natural gas, 9% from nuclear and the remaining 5% from renewable sources such as solar, geothermal, wind, and hydroelectric[1].

Energy demands associated with the industrial-scale production of chemicals and materials is a signicant portion of the United States total energy consumption. Roughly 33% of the energy consumed in the United States is used by the industrial sector, with 22% of this amount used in chemical production, 16% used in petroleum rening, and 14% used in the smelting and rening of metals[1]. Energy costs are also a signicant, if not the most signicant, portion of the total production costs for many materials. For example, roughly 30% of the cost of producing aluminum is the electricity used in the Hall-Heroult process [4], while the cost of energy is roughly 15% of the total cost of producing steel products [5].

The primary sources of power for the industrial sector have been coal, petroleum, hydro-electric and nuclear. The use of alternative energy sources, e.g. solar, wind, geothermal, for industrial materials production has been slow in coming because they are intermittent and low-energy density sources, i.e. no solar power at night, no wind power when the air is still, etc. The success of Concentrating Solar Power (CSP) facilities, such as the 64 MW Nevada Solar-1 (NS-1) near Las Vegas, have demonstrated that solar can have application to high energy density industries such as materials and chemical production.This paper will discuss the current conguration of the NS-1 facility, and potential modications to the facility that will increase both its power efficiency and power density.

Figure 1: Concentrating Solar Power

The Basics of CSP

There are three main divisions within CSP plants: 1) parabolic trough, 2) solar towers, and 3) dish Stirling technology. Of the three, parabolic troughs are considered to be the most commercially ready technology. A parabolic trough CSP plant is basically a traditional steam power plant, running on the Rankine cycle, with one major change, i.e., the coal/gas/nuclear powered steam generator of the traditional power plant has been replaced with a solar field that uses the energy of the sun to generate steam. Concentrating solar thermal power for utility-scale electricity generation is experiencing unprecedented growth. Parabolic trough power plants are considered to be the most commercially ready technology.

Groundwork for commercial parabolic trough power plants was developed by the Luz International Limited from 1984 to 1990. A total of nine solar plants, ranging from 30-80 megawatts electric (MWe) were constructed in California and continue to operate today. The sixth solar electric generating systems (SEGS) plant, SEGS VI has become the focal point of published research on parabolic trough power plants. The design conditions for this study were based on information provided for the 35 MWe SEGS VI plant. In 2007, Nevada Solar One, a 64 MWe parabolic trough power plant, began operations near Las Vegas, Nevada. Acciona Solar Power operates the plant, shown in Figure 2, and it was the first utility-scale parabolic trough power plant built in the new millennium. This plant has been operating well for the past two years.

Construction finished on Andasol 1, shown in Figure 3, in November 2008. This plant is designed with a molten salt storage system capable of 7 hours of full-capacity power production. This is the first commercial parabolic trough plant to implement a molten salt two tank storage system. Thermal storage was utilized in SEGS I but the storage medium was the synthetic oil solar field heat transfer fluid, or HTF. Synthetic oils are no longer considered for a storage medium in part due to their higher cost.

The future of parabolic trough technology is bright as there are over 1000 MW of plants under construction and even more have been announced. Many of these plants claim thermal storage will be integrated into their plant design. The principle advantages of thermal energy storage in a solar parabolic trough power plant are the ability to control the time and quantity of power production. Herrmann asserts thermal storage can be applied for: buffering during transient weather conditions, dispatchability, increased annual capacity factor, and more even distribution of electricity production.

Thermal storage can provide the stability necessary for base load operation and it also can have the economic advantage to discharge surplus power during peak demand hours. Additionally, the annual solar-to-electric efficiency can improve as a result of thermal storage. Price showed that the improvements to turbine start-up, excess heat from the field, improved parasitic losses, and negligible energy loss from below turbine minimum outweigh the storage thermal losses and reduced power plant steam cycle efficiency due to storage.

As seen in Figure 1, concentrating solar power (CSP), uses thousands of mirrors to focus sunlight on solar receivers to capture the consequent heat and use it to create electricity. These mirrors then track the sun to maximize the amount of sunlight available to concentrate. The most well developed of this technology is the parabolic trough, which is used at Nevada Solar One. [6]

Current operating parameters of Nevada Solar-1

Figure 2: Nevada Solar 1

Potential modifications to NS-1

There are a number of ways that the efficiency of CSP plants such as NS-1 can be improved such as incorporation of thermal storage and use of alternative working fluids. A few of those options are the use of storage oil and storage salt.

Thermal storage

The performance of a solar thermal parabolic trough plant with thermal storage is dependent upon the arrangement of the heat exchangers that ultimately transfer energy from the sun into steam. The steam is utilized in a traditional Rankine cycle power plant. The most commercially accepted thermal storage design is an indirect two-tank molten salt storage system where molten salt interacts with the solar field heat transfer fluid (HTF) through a heat exchanger. The molten salt remains in a closed loop with the HTF and the HTF is the heat source for steam generation. An alternate indirect two tank molten salt storage system was proposed where the molten salt was utilized as the heat source for steam generation.The two differing indirect two-tank molten salt storage designs are presented in the sections below.

Storage Design Oil-Water

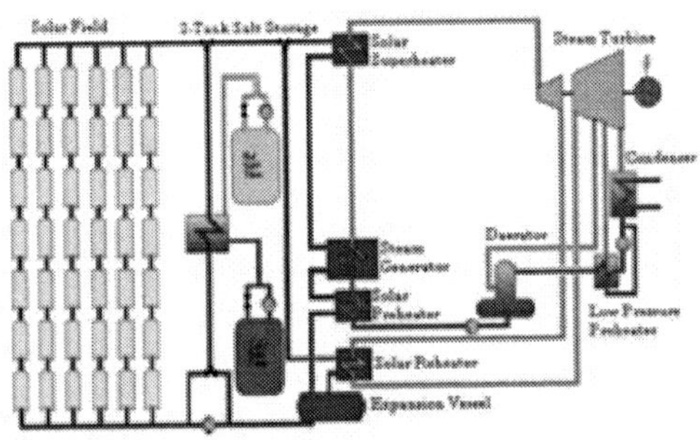

Figure 3: Nevada Solar 1

Indirect two-tank thermal storage can be integrated into a parabolic trough plant. This is the most commercially ready thermal storage design and may be referred to as Storage Oil-Water because the synthetic oil HTF is the heat source in steam generation. While actual operational schemes may be quite complex, the addition of thermal storage does not have to significantly affect the overall operating strategy.The basic operating strategy is to charge thermal storage when the HTF flow rate exceeds the design flow rate for steam generation. Surplus flow travels through the oil-to-salt shell and tube heat exchanger to charge molten salt which exits to the expansion vessel.

During charging, molten salt leaves the cold tank extracts heat from the HTF, and then enters the hot tank. Ideally the design flow rate of HTF is maintained during operating hours so discharging from the hot tank should be performed to maintain the maximum HTF flow rate through the steam train and reheater. Discharging salt from the hot tank to reheat the HTF occurs in the same heat exchanger except flow is reversed. Salt is always maintained on the shell side of the heat exchangers. The first law of thermodynamics requires a temperature drop across a heat exchanger. The temperature of the HTF heated by discharging salt will be lower than the HTF temperature directly from the solar field because the heat has passed through two heat exchangers and an associated heat loss inside the hot tank. This decrease in temperature will result in a decrease in power generation.

The required volume of molten salt is considered to be the volume required to completely fill one tank. Consequently, if one tank is completely filled the other tank is empty. This is a simplification of the actual system, where a minimum volume of salt must be maintained within each tank. When fully charged all the molten salt resides in the hot salt tank at maximum temperature of 386 C. The design temperature of the cold salt tank is 293C. The tanks are considered to be fully mixed thermally and have a heat loss correlation based on surface area of tank. The oil-to-salt heat exchanger must be sized for the discharging capacity of the HTF at the design flow rate for steam generation. Therefore the size of the heat exchangers must be optimized to transfer heat to the design HTF flow rate. If a solar field was designed for 800 W/m2, then a solar multiple of 1.6 with DNI of 1,000 W/m2 would provide double the HTF design flow rate.

For any larger solar multiple, the oil-to-salt heat exchanger area must increase and subsequently its cost will increase. To charge the HTF, the necessary flow rate is withdrawn from the expansion vessel and is mixed with HTF flow from the solar field, if there is any. If there is not enough heat in storage to bring the mass flow rate up to design flow, but enough hot molten salt to generate the minimum amount of power, the hot molten salt is discharged completely. Four MWe was the minimum amount of power assumed necessary for electricity generation. This corresponded to an oil mass flow rate of 40 kg/s. If the HTF flow rate could not reach 40 kg/s, even after thermal storage discharge, the hot salt would dwell in the hot tank and power would not be generated.

Storage Design Salt-Water

The indirect two-tank molten salt storage is referred to as Storage Salt-Water. This is because molten salt is the heat transfer fluid in the steam train and reheater. The synthetic oil is contained in a closed loop around the solar field and the oil-salt heat exchangers. An expansion vessel will still be needed for the solar field. A significant difference between thermal storage plant designs is the number of heat exchangers encountered before transferring heat to the working fluid. No additional heat exchangers are required between the HTF and the working fluid during normal operating conditions for Storage Oil-Water. However, when heat is needed from storage, two additional heat exchangers are utilized.

Heat is first transferred from oil to salt, then from salt back to oil, and finally from oil to the working fluid. The Storage Salt-Water design utilizes two heat exchangers between the synthetic oil and the working fluid for all operations. Storage Salt-Water requires a larger oil-salt heat exchanger area than the Storage Oil-Water case because all of the solar field HTF flow rate must transfer heat to the salt. While the heat exchanger area for Storage Oil-Water does not have to increase, any increase in solar multiple for Storage Salt-Water will result in a larger oil-salt heat exchanger area.

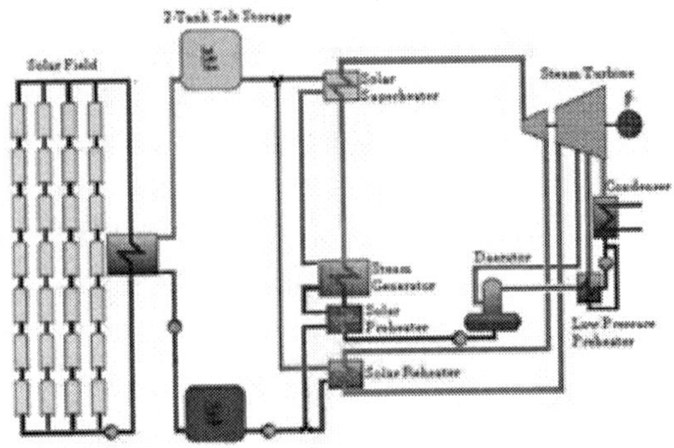

Figure 4: Storage Design Salt-Water

Based on the cost of oil-salt heat exchangers, Storage Oil-Water is heavily favored. Integration of thermal energy storage decreases a plants efficiency for the time period of thermal discharging. This is due to inevitable heat transfer losses to charge the thermal storage medium and also to discharge it. Molten salt is a leading medium for thermal storage and there is discussion it may be circulated through the solar field, thus reducing thermal losses through heat exchange. The description of the molten salt steam generation is also crucial to the analysis of the Storage Salt-Water design.

Working fluids

The synthetic oil used as the solar field heat transfer fluid is a eutectic mixture of diphenyl oxide and biphenyl. Two commercial names for this product are Therminol VP-1 and Dithers A. It is stable up to 399 degrees C. Water boils in the steam generator and exits as a saturated vapor. Due to the latent heat of evaporation the steam generator is the most energy intensive heat exchanger. The superheater utilizes the highest temperature HTF to heat the saturated vapor into superheated steam. The superheated steam performs work on a high pressure turbine and typically loses enough heat to enter the saturation region. The molten salt chosen was a nitrate salt that is composed of 60 % $KNO3$ and 40 % $NaNO3$. Thermal properties for solar salt were found in SAM. Among the benefits of the nitrate salt is its stability up around 600 degrees C. However, a disadvantage is its high melting point; Schulte-Fischedick report that local solidification can occur at 239 degrees C.

When using a salt as a thermal storage medium the temperature of the cold tank is of concern because long periods without charging may lead to freezing conditions. Freezing is a distinct possibility if the salt is not heated by auxiliary heaters. For a plant with a solar multiple close to one, there will be concerns of salt solidification in the winter months, when the flow rate of the field does not exceed design conditions. Freezing would not occur in the normal operations of the Storage Salt-Water case.

Conclusions

Solar power is now becoming a viable commercial competitor. With the addition of thermal storage, solar power can supply power long after the sun has gone down. Solar power also has the least impact on the environment as it does not actively produce pollutants in its operation. This combination of environment-friendly cost-effective power plant makes CSP the best option for our energy needs.

References

[1] DOE/EIA-0384, *Annual Energy Review: 2008.* Washington, DC, June 2009.

[2] EIA, *Annual Energy Review 2008.* Washington, DC, June 2008.

[3] DOE/EIA-0384, *Annual Energy Review: 2008.* Table 1.5. Washington, DC, June 2009.

[4] EIA, *Aluminum Industry Total Energy Use.* Washington, DC, June 2000.

[5] EIA, *Steel Industry Total Energy Use.* Washington, DC, June 2000.

[6] Acciona, *Radiant Solar Energy.* (http://www.acciona-na.com/The-Big-Picture/Sun/Radiant-Solar-Energy.aspx), accessed August 2009.

[7] Acciona, *Nevada Solar One.* (http://www.acciona-na.com/About-Us/Our-Projects/U-S-/Nevada-Solar-One.aspx), accessed August 2009.

Energy Technology 2010
Edited by: Neale R. Neelameggham, Ramana G. Reddy, Cynthia K. Belt, Ann M. Hagni and Subodh Das
TMS (The Minerals, Metals & Materials Society), 2010

RADIO-THERMIONIC INDUCED REACTIONS ON ALUMINUM BASED MATERIALS AND THEIR CORRELATION WITH PLASMON FORMATION AND TRIBOCATALYTIC MECHANISMS

John A. Elton, James Cornwell

Protective Systems, Inc.
7100 Monache Mountain Road, Inyokern, CA 93527

Directed Energy, Magnetron, Subatomic Particles, Scalar Wave, Plasmon, Hydrogen Generation

Abstract

Thermionic induced reactions and radiation resonance catalysis of aluminum and aluminum oxide materials exposed to microwaves and triboemissions will be described and explained from both a fundamental and practical level. Magnetron operation at distinct operating conditions that generates both plasmon and triboemissive radiation will be explained and the resulting tribochemical and radiation resonance catalytic reactions between aluminum, aluminum oxide, water and graphite will be delineated. The generation of hydrogen from these materials using triboemissive radiation and microwaves will be described, and the application of this methodology for hydrogen based technologies will be discussed.

Introduction

<u>Protective Systems Directed Energy Radiation</u>

Protective Systems, Inc, has developed a new radiation device called the EMR (Electromagnetic Resonator) that has unique characteristics that separates it from traditional microwave and rf radiation devices. The radiation from the EMR device is generated using a modified electron-resonance magnetron with proprietary filters and operating characteristics.

<u>Electronic Device Shutoff.</u> When electronic devices such as cell phones, computers, and other IC based electronics are placed within the EMR radiation, these devices become non-functional as long as the EMR beam is active. Upon removal of the EMR radiation the devices either become operational again or become operational after a cold start. The devices are not damaged by the irradiation.

<u>Faraday Cage Interaction Effect.</u> When any of the above electronic devices are placed in a Faraday cage and activated within the Faraday cage without EMR radiation, they operate normally. However, these devices will still be rendered inoperative when an EMR beam generated outside the Faraday cage is directed at the electronic.

<u>Transmission of Scalar Wave Radiation.</u> When a full spectrum antenna is used to characterize the EMR radiation, the only observed radiation is that of the primary microwave generated by the magnetron and multiples of that frequency (2.45 and 4.9 GHz in a 2.45 GHz magnetron). When an EMR device is modified and configured as a receiver and an EMR beam is transmitted to it, a full spectrum of frequencies is generated in the receiver from KHz to GHz frequencies.

Hydrogen Generation

The United States and other countries are attempting to transition into economies that are based upon alternative energy sources not derived from petroleum, coal or natural gas sources. One of the promising new approaches is to generate hydrogen from water and use it as a primary fuel source because it burns clean and provides no pollution byproducts. Protective Systems, Inc. has developed a MicroHygen hydrogen generator that produces hydrogen using EMR radiation.

Theoretical Based Hypotheses

Electron-Resonance Magnetron Operation

Protective System's technology is based upon the theory of operation of electron-resonance magnetrons and additional application of this theory when plasmas, low frequencies and unique operating conditions are employed.

Basic Magnetron Theory [1]

The magnetron is a self-contained microwave oscillator that uses crossed-electron and magnetic fields in the magnetron to produce the high-power output required.

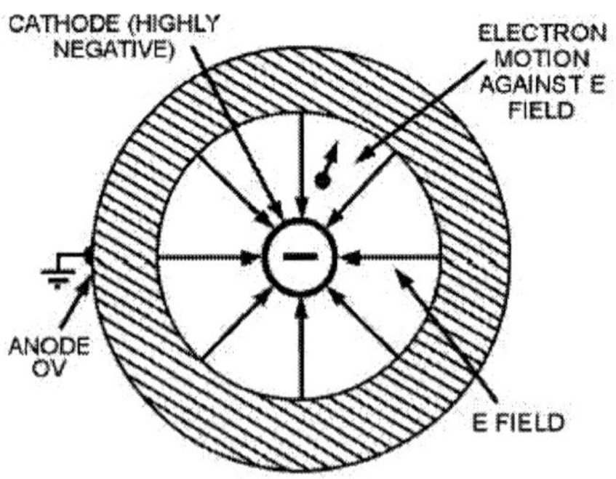

Figure 1. Electron motion in an electric field.

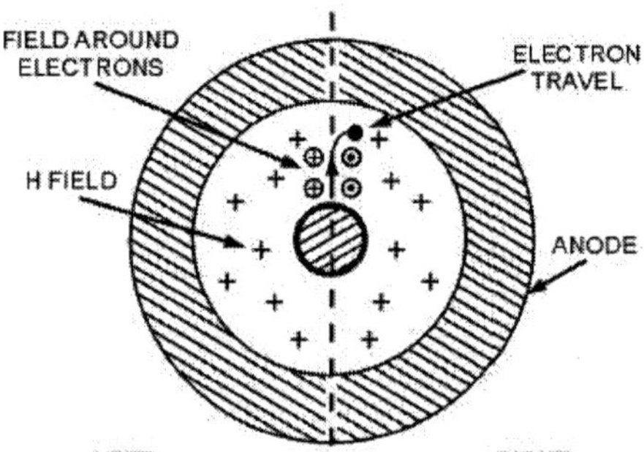

Figure 2. Electron motion in a magnetic field.

Basic Magnetron Operation Magnetron theory of operation is based on the motion of electrons under the influence of combined electric and magnetic fields. This is shown in figure 1 and 2. When no magnetic field exists, heating the cathode results in a uniform and direct movement of the electrons from the cathode to the plate. As the magnetic field surrounding the tube is increased electrons are affected and the electrons proceed to the plate in a curve rather than a direct path. When the magnetron is adjusted appropriately the electrons can be made to orbit the cathode and thereby produce oscillations at microwave frequencies.

Scalar Wave Generation

Typical microwave and RF theory examines the solution to Maxwell's equations when there are appreciable electric and magnetic fields present. An unusual set of field equations and associated radiation exists when Maxwell's equations are solved in the presence of a plasma at low frequencies. These types of waves have some unusual properties as will be discussed below.

Maxwell Equations. Maxwell's equations relate the electric field intensity **E** to the magnetic field intensity **H** and current density **J**. For use in magnetron operation the cylindrical coordinate system is used using ρ, φ and z as the coordinates. For the purposes of magnetron operation, we will look at a solution to Maxwell's equations for all ω but with an assumed time dependence as $\exp(-i\omega t)$, φ dependence as $\exp(ik_\varphi \varphi)$, and z dependence as $\exp(ik_z z)$ [2]. We will then look at the nature of the equations when a plasma is present and when low frequencies are utilized. For the equations to follow we will use as constants the permittivity ε, permeability μ, conductivity σ and resistivity ρ. With these assumptions and definitions Maxwell's equations are as follows:

$$H_\rho = 0 \qquad 1)$$

$$ik_z E_\rho - E_z' = i\omega\mu H_\varphi \qquad 2)$$

$$H_z = 0 \qquad 3)$$

$$-ik_z H_\varphi = (-i\omega\varepsilon + \sigma)E_\rho \qquad 4)$$

$$E_\varphi = 0 \qquad 5)$$

$$(\rho H_\varphi)'/\rho = (-i\omega\varepsilon + \sigma)E_z \qquad 6)$$

The solution of these equations for H_φ is the first order Bessel functions $J_1(k_c\rho)$ and $Y_1(k_c\rho)$ where k_c^2 is defined to be $\omega^2\mu\varepsilon + i\omega\mu\sigma - k_z^2$. What is important to note here is that when $|\omega\varepsilon| << |\sigma|$ and σ is large as in a plasma, the electric field becomes very small but a large magnetic field can exist. Furthermore, since σ for a plasma is imaginary and independent of ω (shown below), the resulting magnetic field will be oscillatory and exist within the plasma. It is also important to note that the wave impedance for this electromagnetic wave is defined as

$$Z_{TM} = E_\rho/H_\varphi = -ik_z/(-i\omega\varepsilon + \sigma) \qquad 7)$$

and for these conditions the impedance becomes quite small. As a result the EMR wave will pass through the plasma with little loss of energy.

At low frequencies other effects are observed. As $\omega \to 0$ in a metal, the skin depth δ can become quite large such that any such wave generated in the magnetron can be propagated through metal walls external to the magnetron. Thus by generating electromagnetic waves at low frequencies in a plasma, electromagnetic waves can be produced that will have essentially no electric field and can pass through metallic barriers. This low frequency condition can be met in such a system because the cutoff frequency is determined by

$$\omega^2\mu\varepsilon + i\omega\mu\sigma - k_z^2 = k_c^2 = 0 \qquad 8)$$

and k_c can be made positive at low frequencies when k_z is imaginary and the wave in the z direction is evanescent. This still provides a wave in the ρ direction that is not evanescent [2].

Plasma Conductivity for a plasma that has negligible collision losses is given by

$$\sigma = iN_eQ_c^2/\omega m_e \tag{9}$$

where N_e is the number of free electrons per cubic meter, Q_c is the charge per electron in coulombs, and m_e is the mass of the electron [3]. It should be noted that this has been derived for a positive traveling wave that depends upon t as exp($-i\omega t$). Use of this σ in Eq. 8 provides a negative real term that modifies the characteristics of the waves produced and provides a mechanism for low frequency scalar waves.

Particle Acceleration

Because the magnetron produces electron motion along the φ direction, plasmon currents can be generated along the φ direction with no E_ρ component to the E field. When such conditions are setup in a magnetron the electrons can be accelerated to velocities approaching the speed of light. These electrons can be used to interact with zero charge subatomic particles as will be shown below.

Idealized Magnetron Theory For the purposes of developing the theory for the magnetron used at Protective Systems, an idealized magnetron was conceptually visualized that has an infinitely long cylindrical plate with a permanent and constant magnetic field strength along the z axis of the cylindrical plate. In addition, the following theory was developed without any assumed anode radius ρ_b at which the E field terminates. This simplifies the mathematics but still approximates the conditions above the cathode in the Protective System's electron-resonance magnetron and approximates the fields that will be produced when a waveguide is attached to the top of the magnetron.

Maxwell Equations We look for a solution to Maxwell's equations in cylindrical coordinates with **E** representing the electric field intensity, and H the magnetic field intensity. Subscripts are used to describe which cylindrical coordinate direction is being utilized (i.e. E_ρ, E_φ, E_z, H_ρ, H_φ, H_z). As before we assume that the solution will have a φ, z and t dependence that goes as $\exp(ik_\varphi\varphi)$, $\exp(ik_z z)$ and $\exp(-i\omega t)$.

Maxwell's equations are then as follows:

$$ik_\varphi E_z/\rho - ik_z E_\varphi = \alpha H_\rho \tag{10}$$

$$-E_z' + ik_z E_\rho = \alpha H_\varphi \tag{11}$$

$$(\rho E_\varphi)'/\rho - ik_\varphi E_\rho/\rho = \alpha H_z \tag{12}$$

$$ik_\varphi H_z/\rho - ik_z H_\varphi = \beta E_\rho \tag{13}$$

$$-H_z' + ik_z H_\rho = \beta E_\varphi \tag{14}$$

$$(\rho H_\varphi)'/\rho - ik_\varphi H_\rho/\rho = \beta E_z \tag{15}$$

where $\alpha = i\omega\mu$ and $\beta = -i\omega\varepsilon + \sigma$.

The solution for these equations are as follows:

$$H_z = [(a_0 (k_c\rho)^{k\varphi})\Sigma_{\kappa=0}^{inf} (-1)^k (k_c\rho)^{2k}/\{2^{2k}k!(k_\varphi+k)!\}] \exp(ik_\varphi\varphi+ik_z z-i\omega t) \quad 16)$$

$$E_z = [(a_1(k_c\rho)^{k\varphi})\Sigma_{\kappa=0}^{inf} (-1)^k (k_c\rho)^{2k}/\{2^{2k}k!(k_\varphi+k)!\}] \exp(ik_\varphi\varphi+ik_z z-i\omega t) \quad 17)$$

$$H_\varphi = (-k_\varphi k_z H_z/\rho - \beta E_z')/(k_c^2) \quad 18)$$

$$E_\varphi = (-k_\varphi k_z E_z/\rho - \alpha H_z')/(k_c^2) \quad 19)$$

$$H_\rho = (- ik_z E_\varphi + ik_\varphi E_z/\rho)/\alpha \quad 20)$$

$$E_\rho = (- ik_z H_\varphi + ik_\varphi H_z/\rho)/\beta \quad 21)$$

where a_0 and a_1 are constants determined by the applied constant electric and magnetic fields imposed upon the magnetron for operation, the associated plasma density generated by these fields and the z axis geometry of the system; and we define $k_c^2 = (\alpha\beta - k_z^2)$.

These solutions have the following implications:

Plasmon and Microwave Generation At a specific ρ and ω determined by the equations

$$\rho_o = k_\varphi/(\alpha\beta - k_z^2)^{0.5} \quad 22)$$

$$\omega^2\mu\varepsilon + i\omega\mu\sigma = 0 \quad 23)$$

and with k_z being imaginary (i.e. evanescent waves in the z direction), electron orbits will be circular and do not return to the cathode or anode. This condition is particularly important to the generation of zero charge subatomic particles because the electrons in circular orbits can both accelerate to high velocities and can interact and affect subatomic particles at this particular ρ value. Associated with the electron currents at ρ_o are the electric and magnetic fields described by the equations above. The microwaves generated at angular frequency ω and the electron currents are strongly correlated due to Maxwell's equations.

Acoustic Wave Generation Because the magnetron is not operated in a vacuum, the electron currents in the φ direction collide and interact with any gas phase or sub-atomic particles present. The electrons thereby accelerate these species up to the velocities that the electrons attain at ρ_o. Since the forces exerted by the electrons are centrifugal in nature at ρ_o, many of these species also attain circular orbits along with the electrons. Finally, because the electron currents have frequency and wavelength characteristics in the φ direction, the gas phase and subatomic particles also inherit these characteristics and an acoustic wave is thereby created that is closely tied to the microwaves generated in the interaction space.

Relativistic Subatomic Particle Wave Function Because the electrons and subatomic particles are approaching velocities near the speed of light, relativistic considerations must be taken into account in deriving any wave functions for particles being accelerated at ρ_o.

The relativistic wave equation for the subatomic particles at ρ_o is determined by first deriving the classical expression describing the movement of a particle being acted upon by a centrifugal force (i.e. the electrons colliding with the subatomic particles at ρ_o) [4, 5]. This expression at ρ_o is given by

$$E = \{(p_\varphi^2 + p_z^2)/2m\} - k/\rho_o \qquad 24)$$

where p_φ is the momentum in the φ direction, p_z is the momentum in the z direction, m is the mass of the particle considered and $-k/\rho_o$ is the potential energy. Note that no electromagnetic components are in Eq. 20 because zero charge particles will not be affected by the electromagnetic force. The corresponding relativistic equation is

$$(E + k/\rho_o)^2 = c^2p^2 + m_o^2c^4 \qquad 25)$$

and replacement with the quantum mechanic operators for E and p gives the relativistic wave equation

$$[-(h^2/4\pi^2)\delta^2/\delta t^2 + \{(2i(h/2\pi)k/\rho_o)\delta/\delta t\} + k^2/\rho_o^2]\psi = [-(h^2c^2/4\pi^2)\nabla^2 + m_o^2c^4]\psi \qquad 26)$$

The solution for this equation is given by

$$\psi = C\exp\{-i(Et/(h/2\pi) + k_\varphi\varphi + k_z z)\} \qquad 27)$$

where C is a constant that will be determined by normalizing the solution over all φ and over z using box normalization or infinity considerations and

$$k_z^2 = \{(E + k/\rho_o)^2 - m_o^2c^4 - (h^2c^2/4\pi^2)k_\varphi^2/\rho_o^2\}/(h^2c^2/4\pi^2) \qquad 28)$$

This equation takes on more relevance when it is written as

$$E_k^2 = \{(E + k/\rho_o)^2 - m_o^2c^4\}/(h^2c^2/4\pi^2) = k_\varphi^2/\rho_o^2 + k_z^2 \qquad 29)$$

and the relationship between kinetic energy E_k and ω ($E_k = h\omega/2\pi$) is taken into account.

$$\omega^2/c^2 = k_\varphi^2/\rho_o^2 + k_z^2 \qquad 30)$$

This is exactly the electromagnetic equation obtained for circular motion (Eq. 22) without the plasma component which has not been taken into account in deriving Eq. 30. It is anticipated that the plasma component will add a term to Eq. 30 equal in magnitude to that in Eq. 22 but interpreted as an interaction between the particles independent of charge. This is presently being derived and will be provided in a future publication. The correspondence between Eq. 22 and Eq. 30 is important because it shows that electromagnetic waves, electrons and zero charge subatomic particles with the same k_φ, k_z, ω and plasma characteristics will have the same kinetic energies, frequencies and energy differences between adjacent states. Under these resonant conditions energy can readily be transferred from one species to another.

Qualitative Application of Theory to EMR Operation

The scalar waves generated by EMR devices are essentially magnetic waves at low frequencies. It is this unique nature of the EMR radiation that allows the radiation to pass through metal enclosures and Faraday cages. The magnetic wave generated by the EMR device can shut off any devices containing integrated circuits due to an effect upon the depletion layer currents within semiconductor junctions.

Electronic Device Shutoff

Figure 3 shows a 310 MHz strobe light used to demonstrate its deactivation when exposed to EMR radiation. Also shown is the remote wireless controller used to activate the strobe device. Figure 4 shows the typical arrangement when performing such a demonstration. When an EMR beam is projected at the strobe light no activation of the strobe light is possible. However, once the EMR device is turned off the strobe light can be activated by the remote control.

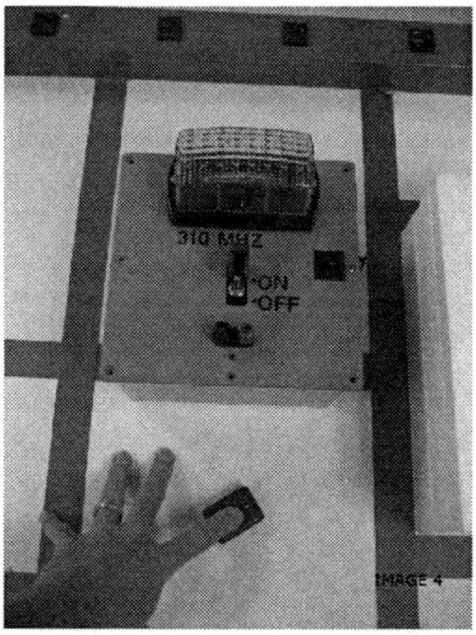

Figure 3. 310 MHz strobe device used to demonstrate electronic device shutoff.

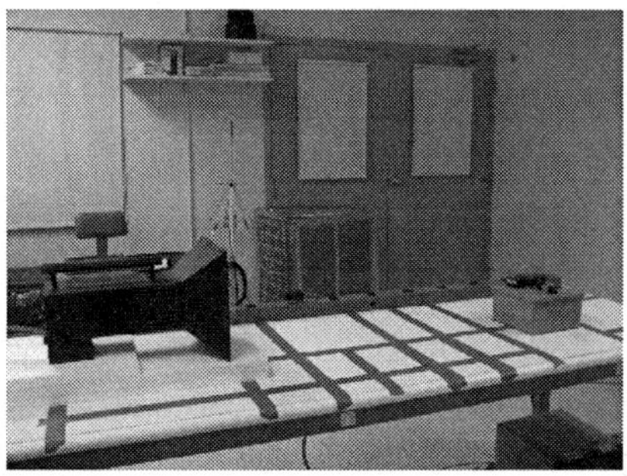

Figure 4. EMR HD-50 in typical configuration demonstrate electronic shutoff using 310 MHz strobe device.

Faraday Cage Interaction Effect

This deactivation of the strobe light and other devices such as cell phones and laptop computers can be performed even when the device is placed in a Faraday cage. Figure 5 shows the activation of the strobe light when no EMR beam is directed at the strobe light external to the Faraday cage. Figure 6 shows the typical configuration of the EMR HD-50 and the Faraday cage with strobe light inside. When the strobe light is activated using the push button on the top of the Faraday cage that is connected to the mechanical linkage that in turn is connected to the remote control inside the Faraday cage, the strobe light will flash until the EMR radiation is turned on and directed at the strobe light. The strobe light will remain off until the EMR radiation is turned off and will resume periodic strobe action at that time.

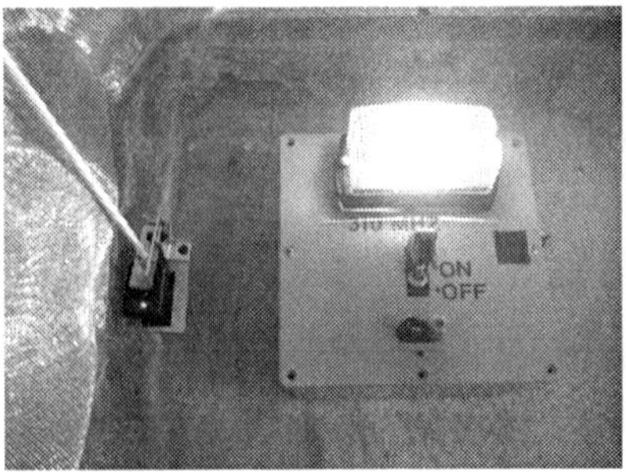

Figure 5. 310 MHz strobe device in closed Faraday cage being activated via mechanical linkage that can be actuated externally.

Figure 6. EMR HD-50 in typical configuration to demonstrate deactivation of 310 MHz strobe device when enclosed in Faraday cage.

Hydrogen Generation from the MicroHygen Generator

The EMR technology has a variety of applications as demonstrated above. For chemical reactions this unique form of energy holds the promise of affecting reaction pathways through generation of different types of radiation (scalar waves, plasmon waves and subatomic particle interactions) and through the selection of specific frequencies and plasma densities. One of our most studied applications of EMR to reaction chemistry is with respect to hydrogen generation.

MicroHygen Apparatus

Protective Systems, Inc. has developed a hydrogen generation system that uses EMR radiation to generate approximately 700 liters/hour of hydrogen. The prototype unit is shown in Figure 7. This system uses a 700 watt 2.45 GHz electron-resonance EMR unit (Figure 7a & b) connected to a reaction chamber (Figure 7a & b) that contains approximately 20 liters of acidified water. The hydrogen generated from this unit is sent through two filters and a mass flow meter (Figure 7b & c) before being provided to a slightly modified propane burner unit (Figure 7d).

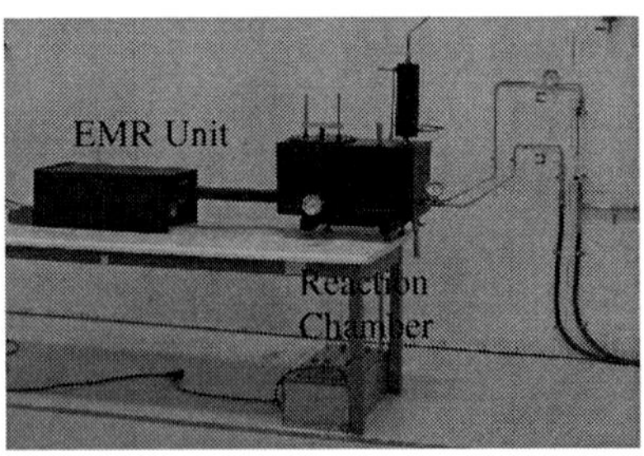

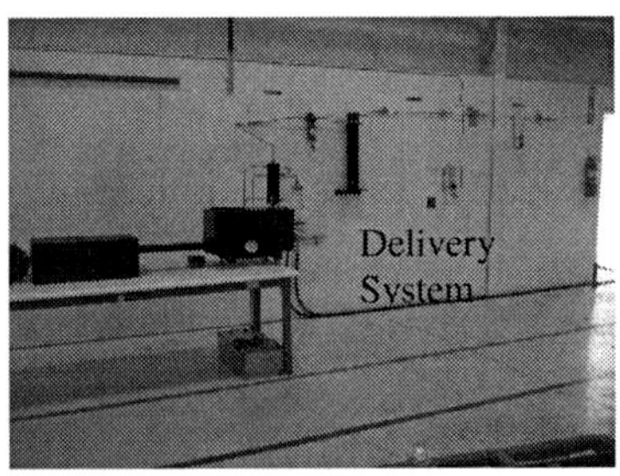

Figure 7a. MicroHygen apparatus showing the EMR unit and the reaction chamber.

Figure 7b. MicroHygen apparatus showing EMR unit, the reaction chamber and the delivery system.

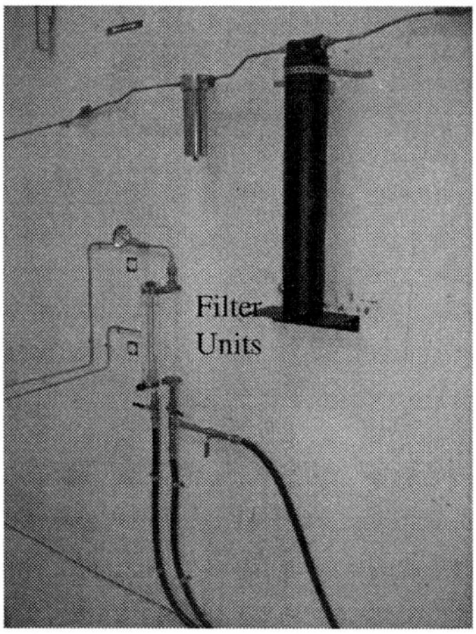

Figure 7c. MicroHygen apparatus showing filter units and delivery system.

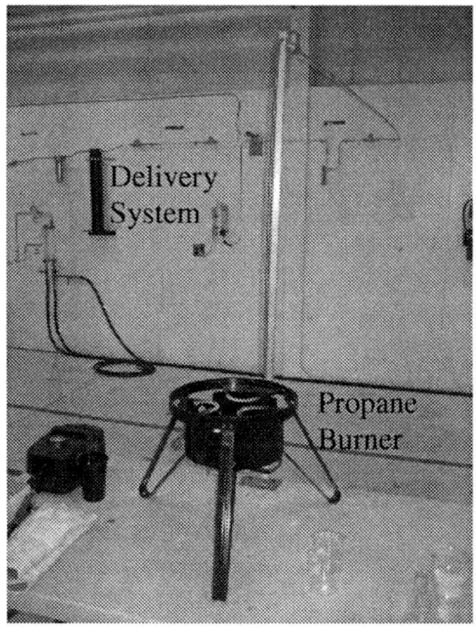

Figure 7d. MicroHygen apparatus showing the delivery system and the propane burner.

Hydrogen Generation with the MicroHygen

The reaction of aluminum with water is utilized in Protective System's MicroHygen technology. We believe that the aluminum/water reaction proceeds rapidly because the EMR radiation disrupts the formation of the aluminum oxide that is generally formed on the surface of aluminum and/or energizes the water molecules or hydronium ions during the reaction.

Basis of Technology The reaction between aluminum, water and acid is used to generate the hydrogen. Initially it was believed that the reaction went according to:

$$2Al + 3H_2O \rightarrow Al_2O_3 + 3H_2 \qquad 31)$$

Recent analysis of the solid reaction byproduct has shown that the reaction proceeds according to the reaction:

$$2Al + 6H_30^+ + 6Cl^- + 6H_2O \rightarrow 2AlCl_3(H_2O)_6 + 3H_2 \qquad 32)$$

This reaction is exothermic with a ΔH^o of -77 kcal/mole H_2. The reaction not only produces heat that can be utilized but also generates the hydrogen needed for a cleaner environment.

Experimental Results Several experiments have been performed with the apparatus shown above. The results of one such experiment are shown in Figures 8 and 9 and Table 1. Figure 8 shows the flow rate of gas produced as a function of time. Figure 9 shows the increase in the temperature of the reaction chamber as a function of time. As noted above the reaction is quite exothermic. The exact composition of the gas has not been determined yet, but the gas burns with a near colorless flame and the results of the byproduct analysis would suggest that the gas is primarily if not completely hydrogen. Water vapor would not be present because the gas is passed through a large desiccant filter before it is burned.

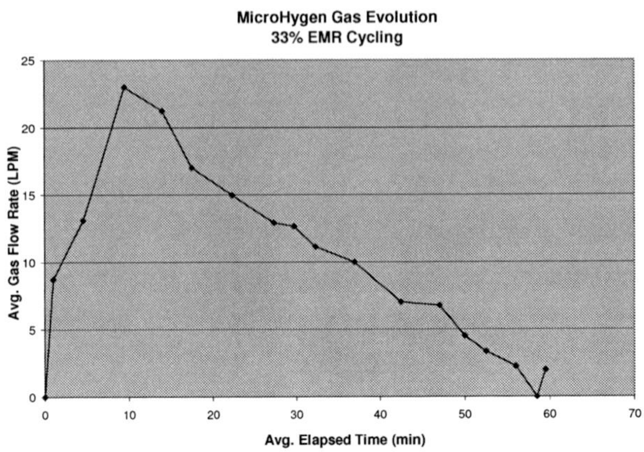

Figure 8. Average gas flow rate observed in the MicroHygen apparatus when the EMR was cycled 5 minutes ON and 10 minutes OFF during the reaction.

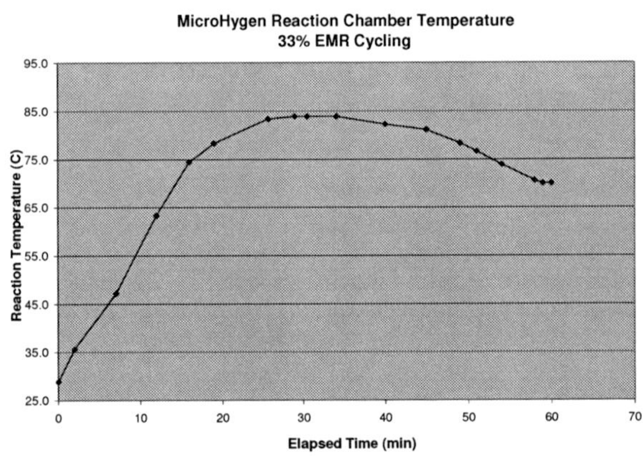

Figure 9. Average reaction chamber temperature observed in the MicroHygen apparatus when the EMR was cycled 5 minutes ON and 10 minutes OFF during the reaction.

Table 1. MicroHygen Experimental Conditions and Results

Experimental Conditions	
EMR Directed Energy Device	
1. Cycle Time %	33
2. On Time (min)	5
3. Off Time (min)	10
Reaction Chamber	
1. Total Volume (liters)	19
2. Weight of Al Rods (g)	1074
Experimental Results	
1. Gas Volume Collected (liters)	687
2. Max. Flow Rate (LPM)	23
3. Al Rod Consumption (g)	712
4. Energy Utilized by Apparatus (KWh)	0.39
5. Energy Utilization Efficiency (%)	92
6. Aluminum Mass Conversion Efficiency (%)	71

Table 1 provides the details of the experiment. The total volume generated in an hour is 687 liters. This will be considerably increased in the future after a subsystem is added to the MicroHygen that will maintain the acidity level so that the reaction rate does not peak and then fall off. In addition this new subsystem will effectively produce hydrogen according to Eq. 31 and therefore provide an approximate 5.6 wt. % of hydrogen in line with the 2010 DOE goal of 6 wt. %. Of note is the fact that a large hydrogen generation rate is obtained with a rather small footprint.

Market Considerations Because the MicroHygen system can utilize scrap aluminum of poor quality, we see this process being used as one of the final stages of aluminum processing when reprocessing of the aluminum scrap is no longer economically viable. Since this low quality aluminum scrap (~$0.07/lb) can be obtained for much less than the commodity price of aluminum (>$1.00/lb), it is a viable alternative to generate heat and hydrogen from this scrap and produce quality brown aluminum oxide (RBAO) as a reaction product.

The worldwide market for RBAO is about 2 million tons per year. The average price presently is about $0.43/lb. This represents a market size of about $2 billion. Thus 1 ton of hydrogen and 17 tons of RBAO can be produced for about $1,900, and the RBAO can be sold for about $14,500 giving a net profit of $12,600/ton H_2.

If the price of hydrogen can be estimated from the cost of converting aluminum hydroxide back to aluminum [6] and an electricity cost of $0.10/KWh is used, then 1 ton of hydrogen would be priced at $12,700. Thus the net profit is almost doubled when the profit from selling the hydrogen is included. Thus from a materials cost of $1,900 a net profit of ~$25,000 can be realized. Of course this crude analysis does not take into account capital, labor and other associated costs involved with any business process.

Summary

The EMR technology introduced herein provides a radiation source with unique characteristics. Because it produces electromagnetic waves with essentially no electric field component, it can penetrate metals and Faraday cages. Its magnetic characteristics provides the ability to deactivate electronic devices with solid state components. The generation of this radiation proceeds in a manner that allows interaction with zero charge subatomic particles and the subsequent inclusion of such particles within the EMR beam. One application of this technology for the generation of hydrogen has been described that utilizes the EMR beam to energize and potentially disrupt the reaction surface.

References

1. "Electrical Engineering Training Series", Integrated Publishing, http://www.tpub.com/content/neets/14183/css/14183_103.htm, pp. 2-23 to 2-38.
2. McIntosh, R. E., Microwave Engineering I notes from ECE 584 class, National Technological University, University of Massachusetts.
3. Lorrain, P. and Corson, D. R., Electromagnetic Fields and Waves, 2nd edition, W. H. Freeman and Company, 1970.
4. Schiff, L. I., Quantum Mechanics, 3rd edition, McGraw-Hill Book Company, 1968.
5. McQuarrie, Donald A., Statistical Mechanics, Harper & Row, 1973, Chapter 1.
6. Petrovic, John and Thomas, George, "Reaction of Aluminum with Water to Produce Hydrogen", Version 1.0, U.S. Department of Energy, 2008.

Patents

The following patents are pertinent to the discussion herein:

1) U.S. Provisional Application No. 60/997,691; "Apparatus and Method for Producing a Continuous Electromagnetic Wave for the Purpose of Inducing Radiation Interference Upon the Surface of Electronic Solid State Conductor", filed November 13, 2007.
2) U.S. Provisional Application No. 60/987,828; "Apparatus and Method for Using a Continuous Electromagnetic Wave to Produce Angular Displacement and Refraction of Emitted Particles", filed November 14, 2007.
3) U.S. Provisional Patent Application 60/532,176; International Patent Application PCT/US05/00078 REF: 1/2005; "Apparatus and Method of Producing Electromagnetic Wave".
4) U.S. Patent Application No. 12,269,876; "Directed-Energy Systems and Methods for Disrupting Electronic Circuits", filed November 12, 2008.

Energy Technology 2010:
Conservation, Greenhouse Gas Reduction and Management, Alternative Energy Sources

Carbon Management and Carbon Dioxide Reduction

Organizers

Subodh Das

Brajendra Mishra

Neale R. Neelameggham

Energy Technology 2010
Edited by: Neale R. Neelameggham, Ramana G. Reddy, Cynthia K. Belt, Ann M. Hagni and Subodh Das
TMS (The Minerals, Metals & Materials Society), 2010

Low Cost Options for Gas Compositional Analysis of Product Streams of Oxy-Combustion Processes with Carbon Capture (an Overview)

John A. Clark, III, Cathy Summers, Danylo Oryshchyn, Stephen Gerdemann, Thomas Ochs, Casey Carney

National Energy Technology Laboratory;
1450 Queen Avenue SW; Albany, Oregon, 97321, USA

Keywords: Oxy-combustion, gas analysis equipment, carbon capture, low cost, CEM

Abstract

Limited analysis of combustion gas streams where carbon capture is not of interest is done mainly to ascertain operation parameters. Stringent regulations on acid gases and particulates promote careful flue gas analysis. Unregulated components are typically not analyzed. Typically not more than four points (two in the boiler, and one each in the flue and the stack) are sampled. A successful oxy-fuel process with carbon capture and compression is anticipated to require as many as 12 sample points with analysis for water, nitrogen, oxygen, argon, NOx, SOx, and carbon monoxide. Carbon dioxide captured for beverages produces a high value product and the cost associated with analysis is warranted. However, carbon capture for disposal must keep all costs to a minimum. This paper will compare alternative approaches for cost-effective sampling and analysis of the carbon dioxide product streams from oxy-fuel combustion, through all stages of production through compression and pipeline/vessel delivery.

Introduction

Analysis of gases in an oxy-combustion process with carbon capture is essential to the operations of the oxy-fired boiler and the subsequent collection, compression, and purification of carbon dioxide. The process is charged with pure oxygen and not-so-pure carbon in the form of coal at the front end, generating a product of mainly carbon dioxide with traces of argon, nitrogen, sulfur compounds, nitrogen compounds, oxygen, hydrogen, and mercury. It is possible that hydrogen chloride and hydrogen sulfide may also be present in trace amounts [1]. The trace components primarily come from impurities contained in coal such as water, sulfur, nitrogen, and volatile organic compounds. Coal also contains a good deal of mineral matter that is transformed into ash during combustion, becoming particulates that can complicate the analysis regimes utilizing standard analytical equipment.

Carbon capture strategies will require that most of the impurities be removed from the carbon dioxide prior to delivery to a pressurized pipeline or containment vessel for transport for sequestration or reuse. Analysis requirements of the captured carbon dioxide will vary depending on transportation system (pipeline) limitations and the planned end use. For example, use as beverage carbonation would require a series of very accurate and carefully referenced compositional analyses to verify the absence of any substance like mercury or sulfur that may have negative health or taste effects, respectively, after the purification stages. If the CO_2 is intended for sequestration, however, contamination with sulfur may be beneficial, as preliminary research has shown it may enhance the rate of mineral carbonation in olivine formations [2]. Also, even down hole injection is the only aim, the presence of O_2 and H_2S in CO_2 cause concern related to the life expectancy of the pipes, pumps, and other equipment used for this purpose due to corrosion [3].

A pulverized coal (pc) boiler facility using air or oxygen that is not concerned with the capture of carbon dioxide might only used limited process and emissions monitoring to maximize efficiency in the boiler and to verify compliance to EPA-regulated components in the flue/stack gas. Typically, this would require four sample locations with limited specific analyses at each port to ensure complete coverage of the gas stream [4]. An oxy-combustion system is anticipated to require better and fuller analysis at more locations during compression and purification stages of the flue gas [5]. Analysis of the combustion, capture, and emission verification gas streams may require as many as 12 sample locations for an oxy-combustion process with carbon capture. One challenge will be to reduce costs of robust gas analysis of such that the continuous emissions monitoring (CEM) equipment deployed for oxy-combustion processes are comparable in price, despite the additional sampling locations, to that of an installation and deployment at a traditional (pc) power plant.

Background and Baseline

Conventional air-fired pc exhaust is substantially different (figure 1) from that of oxy-combustion (figure 2) of coal exhaust [7]. Furthermore, the exhaust of a conventional boiler is typically vented to the atmosphere after scrubbing out only the regulated constituents (SOx, NOx, particulates, etc.), whereas an oxy-combustor would be coupled with a carbon capture scheme that would require the a careful accounting of all combustion product constituents from combustion, collection, all stages of compression, and the final products. Table 1 is based on research conducted by the National Energy Technology Laboratory and its partners on a patented oxy-combustion, carbon capture process called Integrated Pollution Removal (IPR™) [6]. The IPR™ process, which is described elsewhere [8, 9, 10, 11], provides for the integrated removal of heat, contaminates, and pipeline ready carbon dioxide from oxy-combustion processes.

Table 1: Gas products comparison of Air-Fired and Oxy-Combustion with effluent compression

Constituent	Air-fired (traditional)	Oxy-Combustion	Compression stage 1	Compression stage 2	Compression stage 3
			Mass fraction		
H_2O	0.0832	0.3322	0.0695	0.00994	0.0004
Ar	0.0088	0.0115	0.0163	0.0173	0.0175
CO_2	0.1368	0.6132	0.8662	0.9216	0.9305
N_2	0.7342	0.0091	0.0128	0.0136	0.0137
O_2	0.0350	0.0250	0.0353	0.0376	0.0379
SO_2	0.0020	0.0091	0.0000	0.0000	0.0000
CO			Not reported		
NO_2			Not reported		

Table 1 provides as good comparison of the traditional pulverized coal boiler and oxy-coal fired boiler products of combustion and the post combustion process requirements for carbon capture using a 3 stage compression process as an example. Also an emerging concern is the production and mercury which IPR™ is also designed to remove.

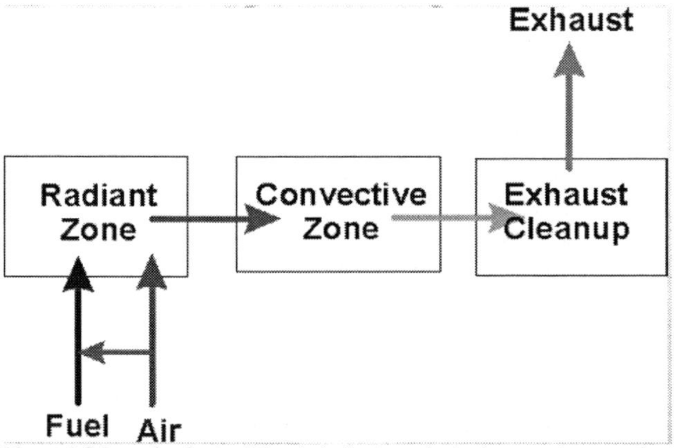

Figure 1 Schematic of a traditional air fired pc plant where most gas product analysis is performed in the exhaust clean-up and exhaust stack per EPA regulation (from [7])

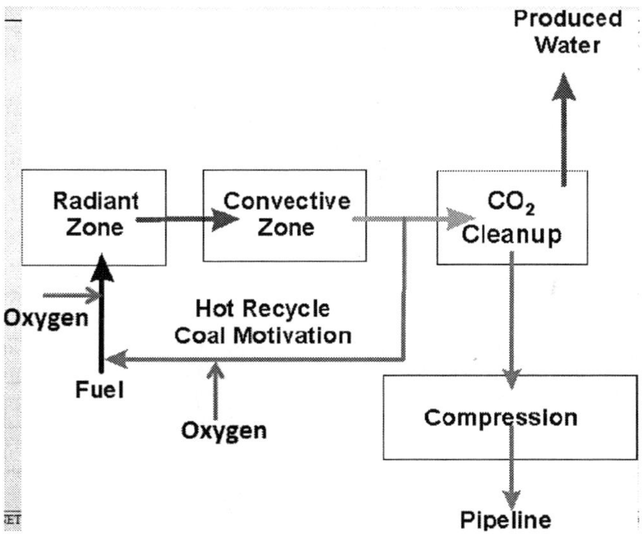

Figure 2 Schematic of Oxy-fuel fired system that would require monitoring of gas constituents in the radiant zone (burners), the recycle line, CO2 clean-up, compression (multiple stages), and pipeline and stack verification (from [7]).

Based on concerns for CEM's listed as drivers for EPRI's Research Program 77 [12], collecting information conducive to careful control an oxy-combustion process with carbon capture and compression, the analyzer(s) should meet the following conditions:

- Real-time analysis of all gas constituents from the combustor and passing through each stage of compression steps
- Accurate, consistent, precise, and highly resolved analysis of all desired constituents
- Commercially available ("off-the-shelf") technology
- High reliability and ease of operation
- Lower cost than traditionally deployed CEM equipment

Starting with the CEM section in the 41st edition of Steam [13], a survey of currently available commercial off-the-shelf equipment was performed to determine potential candidates capable of meeting the requirements listed above. The result of said survey is listed below.

Survey of Commercially Available, Off-the-Shelf Equipment Suitable for Continuous Emissions Monitoring

Gas Chromatography:

Gas chromatography (GC) is an analytical technique that uses columns packed with particles of specific size, composition, and chemical treatment to separate gas species primarily in accordance with their molecular weight. Sampling is done by drawing in samples of the gas to be analyzed into injectors which feed the columns, which separate the gas constituents and feed the detector. The sample is preconditioned to remove moisture and particulate prior to introduction to the injectors. To ensure that the gases remain gas and to facilitate their mobility through the column(s), the injectors and columns are heated to temperatures appropriate for the columns and the gases being analyzed. The separated gases are then passed by a detector, typically a thermo-conductivity detector (TCD), which is calibrated to detect specific gases of interest. Other detectors are available and are used, but the TCD is the most versatile/generic. A carrier gas of helium, argon, or nitrogen is used to facilitate movement of the analyte through the columns and the flush the columns between sample intervals.

Gas chromatography systems packaged for robust industrial environments is a relatively recent development. Those used have typically been deployed in the petrochemical industry. Two vendors were found that market GC equipment suitable for analysis of oxy-combustion gas product analysis. A single GC may be configured to measure all the constituents listed in table 1 with very high resolution over a wide span of compositions for approximately $30,000. Gas chromatography is a reliable technology that is well understood. The biggest drawback is the slow sample analysis time, which can run as long as 5 minutes. Another negative is that GC's are sensitive to water and particulate contamination and optimal performance of a gas conditioning system is essential.

Mass Spectrometry:

Mass Spectrometry (MS) is a technique that sorts the ionized gas constituents in accordance to the ratio mass to atomic charge, and detects the specific compounds in related to specific ratios. The most common deployed of this analytical technique seems to be the "quadrapole" mass spectrometer which makes use of a quadrapole detector to sense the sorted ions of interest. Once restricted to laboratory, this technique has been packaged in industrial and field deployable units. Reportedly mass spectrometers are available that can measure every element from amu 1 through 300 (except those with similar m/z ratios). This covers all combustion products effluent to the oxy-combustion process.

A good MS with capable of measuring a wide range of amu's is anticipated to cost about $100,000. This price may seem a disadvantage, but many mass spectrophotometers can provide analysis over the entire range of constituents in seconds with little purge time required. The MS can analysis for water along with the other products of combustion. A major negative is that it is very sensitive to dusty environments. Also, while the extracted gas need not be dehumidified, the particulate mater (in any) would still need to be filtered.

Tunable Diode Laser Absorption Spectroscopy (TDLAS):

Tuned diode lasers are used to measure characteristics and concentrations of specific molecules in a gas-filled path. This technique may be adapted for in-situ measurements of the high-temperature regions flames to determine region compositional analysis of fuel and oxidants. This technique operates by taking advantage of the fact that gas molecules absorb light at wavelengths specific to the molecule of interest. The laser is used to generate light at unique and discrete wavelengths and a detector measures the difference between the light emitted and the light returned. Typically a single unit can only measure a couple of the desired constituents (e.g. CO and CO_2, SOx and CO, etc.) This analysis can be used to generate information on compositions, temperature, and under controlled conditions pressure.

Some advantages include: real-time, in-process measurement of water vapor up to ~30% is possible with this equipment, instantaneous measurement with good resolution and sensitivity is possible for specific constituents and sample conditioning for either in-situ or extractive measurements in not necessary. The biggest disadvantages are the limited analytical range and the challenges associated with calibration. These units cost in the range of $50,000 – $100,000.

Fourier Transform Infrared:

Fourier Transform Infrared (FTIR) is an analytical technique used to collect infrared patterns that instead indicate absorbance or transmittance of infrared energy measure the interference patterns to develop spectra. FTIR is typically quicker than standard monochromatic IR because all the wavelengths are scanned simultaneously. FTIR is capable of measuring a full range CO, CO_2, CH_4, NO, NO_2, SO_2, and NH_3.

Strengths of this technique include coverage of a broad range of gas constituents, quick analysis times, good accuracy, easy calibration over a wide span, and may be adapted for portability. As with other IR techniques, simple diatomic molecules such as N_2, O_2, and H_2 and monatomic gases He or Ar cannot be measured. Mid-range FTIR's are estimated to cost about $60,000.

Electrochemical Cells:

Electrochemical cell (EC) based gas analyzers operate by measuring the reduction and oxidation potentials of oxygen bearing compounds with the cell's electrolyte. The cell operates similarly to a fuel cell. Component measured with this technique are typically O_2, CO, H_2, SO_2, H_2S.

This is a common technology used by pc power plants to analyze for regulated gas products. The technology is well developed and very reliable. Each unit measures only 2 or 3 constituents, but when coupled with other units, including IR, UV, chemiluminescene, and flame ionization devices, a broad range of analytes can be covered. This technology provided for good accuracy, easy calibration over a wide range of concentrations, and may be adapted for portability.

The major drawback is that the cells have to be replaced every five years and can become contaminated if samples are not properly conditioned to be dry and free of dirt (gas conditioner required). Also, multiple units may be required and each unit may cost between $12,000 and $31,000

<u>Infrared(NIR)/Ultraviolet(UV-Vis) – IR/UV:</u>

Overview - UV-VIS-NIR (190 – 1100 nm) and NIR (900 – 2300 nm) devices measure the absorbance of particular gas components characteristic wavelength. This absorbance is related to the vibrational, rotational, and translational strains of multi-atomic molecules. Detection of noble gases and simple diatomic molecules is not possible with infrared or UV techniques. The NIR instruments seem to be suitable for CO, CO_2 and CH_4, where UV-Vis is suitable for NH_3, NO, NO_2, SO_2 (and H_2S after conversion to SO_2)

This is a common technology used by pc power plants to analyze for regulated gas products. The technology is well developed and very reliable. Each unit measures only 2 or 3 constituents, but when coupled with other units, including electrochemical cell, chemiluminescene, and flame ionization devices, a broad range of analytes can be covered. This technology provided for good accuracy, easy calibration over a wide range of concentrations, and may be adapted for portability.

IR/UV techniques cannot measure simple diatomic molecules such as N_2, O_2, and H_2 nor atomic gases He or Ar. Also, multiple units may be required and each unit may cost between $12,000 and $31,000

<u>Chemiluminescence:</u>

Chemiluminescence (Chem) based detector use the phenomenon that certain excited gases release light and heat as they move from a more energetic state to a less energetic state. This type of detector is usually limited to the detection of NO/NO_2 but some models have been adapted to measure O_2 as well.

This is a common technology used by pc power plants to analyze for regulated gas products. The technology is well developed and very reliable. Although this technique is used for NO/NO_2, but when coupled with other units, including IR, UV, electrochemical cells, and flame ionization devices, a broad range of analytes can be covered. This technology provided for good accuracy, easy calibration over a wide range of concentrations, and may be adapted for portability.

Chemiluminescence has high sensitivity to NOx compounds, but is not used for analysis of other compounds. For complete analysis, this technique must be coupled with other techniques, as with the EC and IR/UV, and each unit may cost about $12,000.

Flame ionization detection:

Flame ionization detectors (FID) are used to detect hydrocarbons by using a hydrogen flame to ionize the hydrocarbon species coupled with a detector (collector) sensitive to ionized carbon and hydrogen. The ions hitting the sensor generate current that corresponds to mass of the ions. This is not volumetric and the carrier gas is less important with this technique. When the flame ionization detector is coupled to a chromatography column, the detector yields the mass of the sorted species. If the detector is not connected to a GC, then the resultant measurement simply relates to the total hydrocarbons present. In this case, the total hydrocarbon may be reported on a methane basis. Flame ionizations detectors destroy the sample they are detecting and hence is usually the last stage of any analytical scheme. This detector is common on GCs and in hand held/portable continuous gas analyzers.

This is a common technology used by pc power plants to analyze for regulated gas products. The technology is well developed and very reliable. Although this technique is used for NO/NO_2, but when coupled with other units, including IR, UV, electrochemical cells, and flame ionization devices, a broad range of analytes can be covered. This technology provided for good accuracy, easy calibration over a wide range of concentrations, and may be adapted for portability.

FID's are great because they are sensitive to only hydrocarbons and are typically coupled with either GC's or IR/UV/EC equipment. This technique destroys the sample, and is used only after other techniques are applied. FID detector prices vary widely and depend strongly on how they are packaged.

Other Instrumentation:

At the time of the survey, other techniques such as gas analysis using Raman spectrometry [14] and in-situ FTIR [15] are theoretically possible and are currently in the development phase, but there are no "off-the-shelf" units currently commercial ready for continuous gas analysis applications. Once fully developed, new techniques should also be considered as possible "low cost" options to provide gas analysis for oxy-combustion systems utilizing carbon capture.

Table 2: Summary of Commercially Available Gas Analysis Equipment for use as CEM

Technique	Wide range of elements/compounds measured simultaneously	Rapid analysis/response time	May be coupled with other techniques	Conditioned gas required	Robust in dirty environments	Good resolution	Rapid scan time
GC	yes	no	yes	yes	available	yes	no
FTIR	no	yes	yes	some	available	yes	yes
MS	yes	yes	yes	some	no	yes	yes
TDLAS	no	yes	no	no	yes	yes	ses
EC	no	yes	yes	yes	no	yes	yes
IR/UV	no	yes	yes	yes	no	yes	yes
Chem	no	yes	yes	yes	no	yes	yes
FID	no	yes	yes	yes	no	yes	yes

Strategies for "Low Cost" Arrangements for Twelve Sample Ports

Oxy-combustion processes with carbon capture will require analysis from multiple sample ports, perhaps as many as 12, to ensure that combustion, flue gas recirculation, and capture processes are functioning adequately. Using the NETL patented IPR™ carbon capture process (figure 3) [9] as an example, there are 6 gas sample ports. They are at the 1) spray tower inlet, 2) stage 1 compression inlet, 3) stage 2 compression inlet, 4) stage 3 compression inlet, 5) stage 4 compression inlet, and 6) the final de-hydrated compressed carbon dioxide product. This example only deals with the capture and compression of the oxy-combustion product and by-products. CEM devices would also be required to monitor the performance of the boiler side of the operation as well to monitor boiler performance and flue gas recirculation during heat operations.

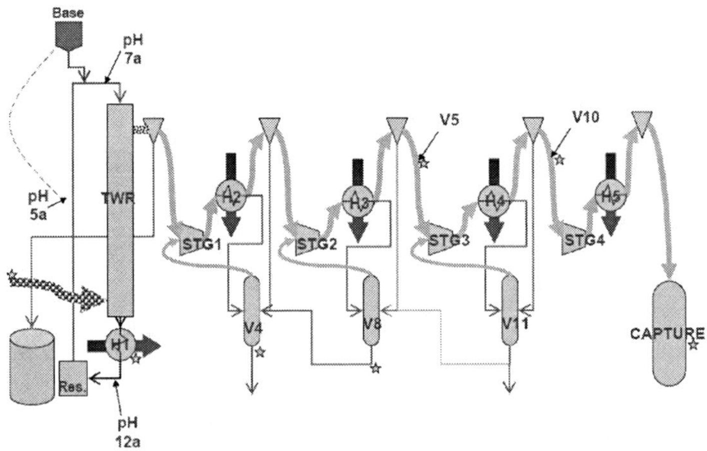

Figure 3: Simplified concept diagram of IPR™ process for the removal of carbon dioxide and pollutants [9]

The number of gas sample ports for an oxy-combustion process is anticipated to vary with the size and geometry of the boiler and the number and distribution of burners, perhaps as many as 6. This may be different from the strategy for traditional air fired boilers because only limited data has been taken to monitor combustion performance in the boiler where CO and O_2 may be monitored in a couple of locations to give indication of the completeness of combustion. The primary focus for traditional air-fired pc boilers is on EPA requirements on SO_x, NO_x, O_2, and CO_2 emissions monitoring and reporting from the stack [16]. To avoid issues with stratification of gas in the stack, up to 4 sample ports are required to cover the entire stack diameter [4]. For the oxy-combustion process utilizing FGR, research [5] has indicated that more careful and thorough monitoring of the combustion process and recycle ratio will lead to enhanced system performance. Therefore, we might anticipate sample ports in the FGR line and in several part of the boiler to monitor for CO_2, CO, SO_x, NO_x, O_2, N_2, and Ar as an indicator of performance and to detect air ingress.

All totaled, even without a stack (this is a capture process), the number of gas sample ports for an oxy-combustion process may equal as many as 12. The potential CEM techniques are listed above, however sampling may be done in-situ, full extractive dry, full extractive wet, and dilution extractive [17]. For the purpose of our analysis and based on our operational experience with CEMs, we will consider in-situ and full extractive dry technologies as the most viable at this time because these are the most commonly accepted practices in current power plant emissions monitoring programs. In-situ analysis will require little sample preparation. For the dry extractive techniques, the gases will need to be conditioned (filtered and de-watered) prior to entering the analyzer. For example, the LDS Analyzer is reportedly insensitive to both water and particulates (within reason – water must be vapor and opacity must permit penetration of laser beam), whereas gas chromatographs and mass spectrophotometers are especially sensitive to particulate contamination and the GC is particularly sensitive to condensable moisture. Gas conditioning systems range in cost from $3500 - $10,000. Multiple vendors exist. Gas conditioning systems can also be engineered and assembled in-house.

Another consideration is gas switching valve systems that allow multiple gas sample lines to share the same analyzers. Still another option is selection of single analyzers that accept multiple probes that can be sampled at intervals that meet emission monitoring and analytical sampling requirements. CEM sharing options (CEM multiplexing) EPA as acceptable methods of sampling as long as, for the current stack emission verification programs, the samples monitoting frequencies are maintained for each regulated constituent [4]. Each sample valve is anticipated to add approximately $5000 to the sampling system cost. The cost of tubing, ancillaries (pumps, wiring, etc.), and the data acquisition system were not considered at this point.

Two possible scenarios of many that may exist that demonstrate the broad range of costs associated with effective analysis of gas components from multiple sample ports are as follows:

Scenario One - (Twelve Sample Ports with a Traditional Extractive Dry Approach using standard CEMs):

Considering a traditional approach, such as that presented by J. Ron Jernigan of Thermo Electron [18], currently used by air-fired pulverized coal plants to verify emission requirements, each port is connected to a bank of analyzers to determine SOx, NOx, O_2 (or CO), and CO_2. Determine the lowest cost analyzers for each particular port and deploy an analytical rack to each port to measure specifically the components of interest. This is the approach supported in the EPA guidelines and is what is done in power plants currently. Utilities limit the number of sample ports to control system cost.

In this case, each sample port will require a gas conditioning system. At an assumed average cost per system of $5000, the cost for gas conditioning will be 12 x $5000 = $60,000.

To cover the analysis of the specified components using standard packaged IR, UV, Electrochemical Cell, and Chemiluminescence techniques, three separate units are anticipated at each sample port:

Analyzers for all listed components: 12 x 3 x $15,000 = $540,000

Estimated total for traditional deployment for 12 sample ports: $600,000

Scenario Two - (Twelve Sample Ports connected to a high-speed Mass Spectrometer via Single Gas Conditioner and Gas Switching Valve):

This approach would require investment in a rapid, flexible (broad range) analyzer and multiplex the samples past the analyzer to measure gas in a reasonable timeframe from each sample port.

Only a single gas conditioner is required in this scenario:
Gas conditioning system: 1x$5000 = $5000

The cost to plumb 12 ports into a single switching valve:
Multiplexing sample valve is 1x$5000=$5000

Analysis of all species of interest can be achieved using MS:
Mass Spectrometer 1x$70,000=$70,000

Estimated total for multiplexed sample deployment of Mass Spectroscopy for 12 sample ports:
$80,000

Of course, both scenarios represent the extremes and combinations of each might as well prove suitable for other "lower cost" solutions for multi-component gas analysis from multiple sample ports.

Summary

Oxy-combustion has significantly different monitoring requirements than from traditional pc gas component analysis. For proper operation, more frequent and more thorough analysis of the gas streams from each the combustion, recirculation, and the carbon capture sections. Emissions monitoring equipment used for gas analysis should meet the following conditions:

- Real-time analysis of all gas constituents from the combustor and passing through each stage of compression steps
- Accurate, consistent, precise, and highly resolved analysis of all desired constituents
- Commercially available ("off-the-shelf") technology
- High reliability and ease of operation
- Lower cost than traditionally deployed CEM equipment

In our survey, in addition to the traditional CEM analyzers based on the electrochemical cell (EC), Infrared/Ultraviolet (IR/UV), and Chemiluminescence (Chem), and Flame Ionization Detector (FID) technologies, the Gas Chromatography (GC), Mass Spectrometry(MS), and Fourier Transform Infrared (FTIR) may be suitable techniques for the more robust requirements of oxy-combustion systems. Also, Tunable Diode Laser Absorption Spectroscopy (TDLAS) was noted as an emerging robust in-situ technique capable of limited compositional analysis.

MS was noted as having the broadest range of analytical capability. As such, mass spectroscopy would allow for the analysis of all the traditional components (water, carbon dioxide, oxygen, argon, nitrogen, sulfur dioxide, nitrogen dioxide, etc.) and the detection and verification of the presence or lack thereof of gas constituents like H_2S, HCl, etc. that may have only been anticipated previously.

Using a single robust, fast, full-range analyzer and multiplexing the ports has huge potential to reduce costs if a suitable analyzer of sufficient resolution and speed can be sourced. The most likely candidate for this application is mass spectroscopy because of its speed, range, and resolution. Although the capital investment in a MS unit is pricey, the potential of using the unit for multiple analysis streams on multiple ports has the potential to reduce the overall system costs.

References

[1] Vince White et al., *Purification of Oxyfuel-Derived CO_2*, Energy Procedia, 1 (2009), 399-406.

[2] Cathy Summers, Dave Dahlin, and Thomas Ochs, *The Effect of SO_2 on Mineral Carbonation in Batch Tests,* Proceedings of the Twenty Ninth International Technical Conference on Coal Utilization & Fuel Systems; Coal Technology Association, Gaithersburg, MD, 2004.

[3] Cal Cooper, *A technical basis for carbon dioxide storage*, Energy Procedia, 1 (2009), 1727- 1733.

[4] "Plain English Guide to the Part 75 Rule", (U.S. Environmental Protection Agency, Clean Air Markets Division, Washington, DC, June 2009).

[5] Raymond Tan et al., "Oxy-Coal Combustion with Flue Gas Recycle For the Power Generation Industry – A literature review", (Study Report IFRF Doc. No. G 23/y/1, International Flame Research Foundation, Valsen Noord, September 2005).

[6] E. David Huckaby, "U.S. DOE/NETL Office of Research and Development – Oxyfuel Combustion Progress Update," (Presented at the Annual NETL CO2 Capture Technology for Existing Plants R&D Meeting, Pittsburgh, PA), March 2009.

[7] Brian Patrick, "Jupiter Oxygen: Annual NETL CO2 Capture Technology for Existing Plants R&D Meeting", (presented at the Annual NETL CO2 Capture Technology for Existing Plants R&D Meeting, U.S. Department of Energy, National Energy Technology Laboratory, March 2009)

[8] Tom Ochs et al., *Oxy-fuel Combustion and Integrated Pollution Removal as Retrofit Technologies for Removing CO2 from Coal Fired Power Plants*, PowerGen International, (2007).

[9] Tom Ochs et al., *Oxy-Natural Gas Firing of the Jupiter Oxygen Oxy-Fuel Test Facility*, Proceedings of the 34th International Conference on Coal Utilization and Fuel Systems; Clearwater, FL, June 2009.

[10] Danylo Oryshchyn et al.,*Capture by Integrated Pollution Removal – Thermal and Process Integration for Delivery of Sequestration-Ready CO_2*, NETL (2008)

[11] S. Gerdemann et al., Developments in Integrated Pollutant Removal for Low-Emission Oxy-Fuel Combustion, DOE/ARC-2005-026.

[12] George Offen, program manager, "Continuous Emissions Monitoring – Program 77", Electric Power Research Institute (EPRI) (2009), (www.my.epri.com).

[13] John B. Kitto and Steven C. Stultz, eds., *Steam/its generation and use. 41st edition* (Barbarton, OH (USA), 2005),

[14] Black and Veatch, Lawrence F. Drbal, ed., *Power Plant Engineering* (New York, NY: Chapman and Hall, 1996), Ch 14.

[15] P. Jackson et al., *In situ Fourier Transform-Infrared (FT-IR) analysis of carbon dioxide absorption and desorption in amine solutions*, Energy Procedia, 1 (2009), 985-994.

[16] J. Kiefer et al., *Design and characterization of a Raman-scattering-based sensor system for temporally resolved gas analysis and its application in a gas turbine power plant, Measuement Science Technology*, 19 (2008), 085408 (9pp).

[17] Thomas C. Elliot, Koa Chen, and Robert C. Swanekamp, *Standard Handbook of Powerplant Engineering* (McGraw-Hill, New York, 1998), .

[18] J. Ron Jernigan, "An Overview of the Current NO_x, O_2, SO_2 Analyzers and Sample Acquisition Technologies and Equipment Installed by the Electric Utility Industry", (Paper presented at the 2nd U.S.-China NO_x and SO_2 Control Workshop, Dalian Bangchui Island Hotel, Zhongshan District, Dalian, Liaoning Province, The People's Republic of China), August 2005.

Energy Technology 2010
Edited by: Neale R. Neelameggham, Ramana G. Reddy, Cynthia K. Belt, Ann M. Hagni and Subodh Das
TMS (The Minerals, Metals & Materials Society), 2010

STRATEGIC APPROACHES FOR CO_2 REDUCTION RATE FROM FOSSIL FUEL USE IN STEEL INDUSTRY

Malti Goel

Former Advisor, Ministry of Science & Technology, Govt. of India
and CSIR Emeritus Scientist, Jawahar Lal Nehru
University, New Delhi – 110067, India

Key words: CO_2 emissions; Coal utilization, Steel, India, Strategy

Abstract

Technology advancements in fossil fuel based energy generation and consumption are currently being aimed at minimization of CO_2 emissions. To attain energy security, India must accelerate the pace of economic development, while achieving reduction in CO_2 emissions per unit of energy consumed. It would need access to advanced coal utilization technologies to be demonstrated on indigenous coal characteristics. Of specific importance would be: improvement in coal quality, increasing energy efficiency and suitability of other CO_2 sequestration technologies for reducing CO_2 footprints. There is need to address related risks and financial burden through adoption of policies and regulations including the issue of open access in energy research. In this paper, seven important areas of strategic plan that need to be pursued for managing CO_2 emissions in steel energy sector in India are discussed.

Introduction

World is facing a crisis of climate change, which follows in the footsteps of global energy and environment crises. Accumulation of carbon dioxide from the generation and utilization of energy is leading to greenhouse warming and climate change. The global warming is causing the boundary between four seasons to be fuzzy and resulting in more spiky weather. Significance of its impact on environment is also determining the technology development needs in various sectors of economy for taking remedial measures.

To respond to climate change concerns, the steel industry is undergoing a transition. It is at the threshold of a rapid change in technology as it targets innovation to achieve 50% reduction of CO_2 emissions per ton of steel produced. Fossil fuel use is the main contributor of increasing concentration of carbon dioxide in the atmosphere. The control of air pollution and CO_2 capture are thus relevant for a coal based industry, whether thermal power plant or a demand side energy intensive industry. World steel production has shown phenomenal increase in first eight years of 21st century increasing from 780 Mt in 2000 to 1343 Mt in 2008 [1]. It was merely 28 Mt in 1900. The global CO_2 emissions from steel industry are also increasing and became 2550 Mt (3-4% of total GHG). Indian industry contribution was 161 Mt (15% of direct emissions from industry). It is likely to grow further to keep pace with the accelerated economic growth of 8-9% per annum.

Growing Steel Production in India

Iron has been backbone of our modern civilization. Looking back, the period from 600 - 50 B.C. is known as *Iron Age*. Indian craftsmanship achieved a high degree of perfection in wrought iron making as early as 400 B.C. An example is historical monument of Qutub Minar built in 375A.D. This famous iron pillar was built at Vishnu Temple of Mathura, from where it was transported to Delhi in 1200 A.D. when the Qutub Minar was built in New Delhi. This engineering marvel has not shown any sign of corrosion for more than 1600 years of its existence. The technology of ultra high carbon steel in small crucibles got established in the medieval period and spread throughout the world from India. After a slowdown, the independent India embarked on the growth of modern iron & steel industry beginning in 1951.

The New Industrial policy 1991 in liberalization regime has given considerable boost to iron & steel industry. From 14 Mt of steel produced in 1991, the production doubled to 28 Mt in the year 2000 and nearly quadrupled to 54 Mt in 2008. The growth of steel production since 1951 is shown in Figure 1. National Steel Policy 2005 targeted steel production as 110 M tons in 2019-20. These targets seemed ambitious, but later increased to 200 Mt. It is evident from current trends that the likely production in 2020 is expected to become 293 Mt.

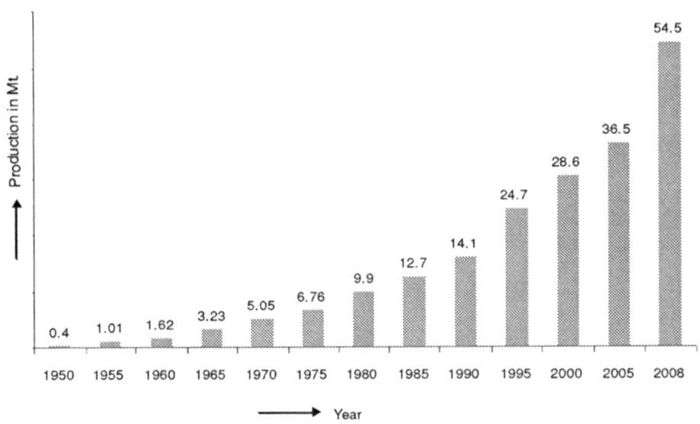

Figure 1. Growth of steel production since 1951

In last 20 years, from nowhere, India has emerged largest producer of coal based Directly Reduced Iron (DRI) technology. Currently, India holds 7th position in steel production. Indian steel industry mainly comprises of large integrated and non integrated steel plants representing 4% of the global production and 4.5% CO_2 emissions. Research & Development (R&D) has been taken up in a mission mode through a high level Empowered Committee (EC). The committee set up by Government of India has played a proactive role in providing overall direction to research efforts in the country. It has been approving research projects for full or joint support with industry or other ministry's participation. Innovative research partnerships among different type of institutions have been facilitated [2]. Ore preparation, beneficiation of iron ore slimes, process development, improving blast furnace productivity, reducing cost of manufacture, development of new products, nano-crystalline iron powders from blue dust, optimization of coal use, production of different grades of steel are some of the research initiatives taken by the Indian industry in the last five years [3]. Human resource development has also been targeted in the mission mode and centers of excellence are proposed to come up in technology institutes.

Over a dozen public and private players are engaged in steel manufacturing to reach world standards. Yet there is scope for improving ecology further. The areas in which technological advancements has taken place or required to take place are; enhancing energy efficiency in steel production, improving quality of raw material, utilization of waste from steel industry as well as knowledge transfer and sharing of new technologies among others [4]. In the following we address seven important areas of strategic plan that needs to be pursued in managing CO_2 emissions in steel energy sector in India.

I. Reducing Specific Energy Consumption in Steel Industry

Most critical factor for the faster growth of steel industry is reducing specific energy consumption. Energy consumption is directly related to CO_2 emissions. Steel industry is the largest consumer of energy among all industrial sectors. It consumes about 10% of total electricity and 27% of coal used by Indian industry. Energy cost contributes nearly 30-35% of this sector's production cost, coking coal accounts for 65-85% of primary energy source. The technology up-gradation measures so far have resulted in reduction of specific energy consumptions by 25% since 1980. In the last 10 years it has fallen from 8.7 Gcal / t cs to 6.7 Gcal / t cs. In 2007-08 world specific consumption was 4.5 Gcal / t cs (Figure 2). Major components of an integrated steel plant contributing to energy consumption and CO_2 emissions (both direct and indirect) are coke oven, sinter plant, basic oxygen furnace and rolling mills, besides the blast furnace. Technologies need to be identified for environment accounting and carbon accounting. Reduction in energy consumption up to 500 million Gcal /year is possible.

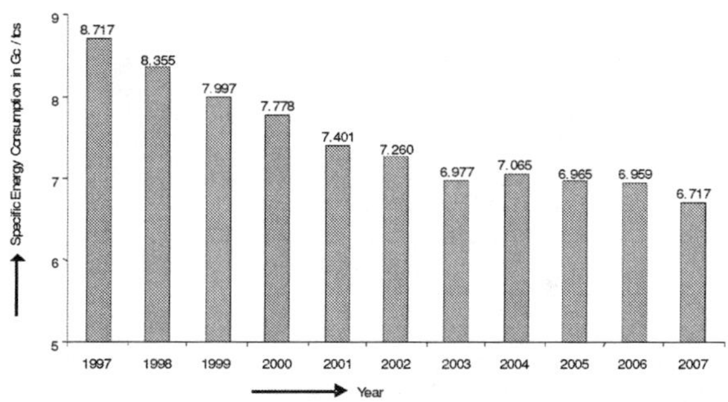

Figure 2. Specific energy consumption in crude steel production (ref [5])

India has taken several policy measures for reducing specific energy consumption in steel making process. Charter on Corporate Responsibility for Environment Protection (CREP) came in to being in 2003. For steel industries it suggests adoption of clean and efficient technologies.

II. Access to Beneficiated Coal for Industry

Coal is primarily consumed in the steel industry through the blast furnace route that accounts for almost 75% of the world coke production. Access to advanced coal utilization technology is of vital importance for the growth of industry and its improved environmental performance.

Technology of Coal Beneficiation

Indian steel industry is the main user of indigenous coking coal resources. Due to its high ash content tests on washability of coal started way back in 1920s. Tata Iron and Steel Co. was first to set up two coal washeries in 1951. In view of criticality of coal for the growth of steel industry, it has in past received tremendous policy support at the national level for access to coking coal. Various technical problems associated with beneficiation of difficult to wash coal have been addressed from time to time and optimization of recoveries targeted [6]. The specifications of Blast Furnace (BF) coke in different countries are depicted in Table I. Except for ash content, other parameters are similar. The problem of ash is being tackled in the recent years by blending it with imported low ash coal. Application of emerging technologies like Oleo-flotation, Oil agglomeration, Column Flotation and Vorsyl Separator for difficult to wash coal has been attempted in R&D mode. Field trials have been conducted at existing washeries. In view of limitation of physical methods of reducing ash content, chemical methods have been tried. Initial trials of chemical cleaning of high mineral content in coal have revealed that solvent refining method is not exactly beneficial for the steel making.

Table I. Country wise specifications of blast furnace coke – source, ref [6]

Parameters	UK	W. Germany	U.S.S.R	Poland	Japan	India (ISI)
Moisture	<3	<5	<5-	<6	3-4	4 (max)
Volatile matter	-	-	1.4-1.0	-	-	1.5 (max)
Ash%	<8	<9.8-10.2	<10-12	<11.5-12.5	10-12	25 (max)
Sulphur %	<0.0	0.9-1.2	1.79-2.0	-	0.6	0.7
Size (mm)	20-63	40-80 25-30	40-30	>40	15-75	-
Micum : M40 Index : M10	>75 7	>34	>73-80 8-9	>63-69 8-10	>75-80 -	78 (max) 10 (max)

Meeting Coking Coal Demand

To enhance the indigenous utilization of coking coal Tata Steel has demonstrated Stamp Charging of low volatile medium coking coal to improve upon the coke quality. Stamp charging is done for optimization of bulk density of coal cake for increasing the productivity improves coke quality. Stamp charging offers environmental superiority and economic benefits in comparison with conventional top charging, because the charging gases are largely eliminated. In this process, visible fugitive emissions (charging emissions, leaking doors, lids and off-gases) are considerably reduced. Mathematic model to select suitable coal blend for stamp charged coal ovens and also to predict coke oven gas composition its heating value and amount of power that can be generated have been developed at Tata Steel. The technology has helped Tata Steel to become one of the lowest cost steel producers in the world.

Strategy for CO_2 reduction from fossil fuel use demands easy access to beneficiated coal. It requires technological challenges to be met for coal beneficiation as well as coal mining. Interior raw material such as high ash content of coking coal even after blending, results in higher energy consumption and higher CO_2 emissions. Further improvement in coal quality and making North-East coal having high volatile content, low ash and high sulphur amenable to use is still a challenge.

Using Non-coking Coal

In view of limited coking coal resource, non coking coal based Direct Reduced Iron (DRI) has been promoted. The COREX technology adopted by Jindal Vijaynagar Steel Ltd. (JVSL) in 2003 uses non-coking coal (as CO and H_2) for reduction of iron oxide by transforming the iron oxide to metallic iron; calcinations of lime stone and dolomite and carbon deposition reaction and formation of Fe_3C [7]. Reduction of iron oxide by H_2 and calcinations reactions is endothermic. On the other hand, reduction of iron oxide by CO gas and carbon deposition reactions is exothermic in nature. After a residence time of about 6 hrs inside the shaft the iron bearing material gets reduced to over 95% metallization. Subsequently discharge screws convey the directly reduced iron from the reduction shaft, to the melter-gasifier. Over 300 small and big units of DRI are in operation / pipe line. Coal based units are located in the coal belts in Orrisa, Chattisgarth, Jaharkhand and West Bengal. The challenge lies in identifying non-coking coal resources, which can produce sponge iron grade coal for indigenous coal utilization.

III. Major Thrust on Waste Recycling

Recycling significantly reduces energy consumption. In the total steel production, 40% is met from recycling of scrap steel worldwide. Steel scrap is 100% recyclable. Use of Iron bearing galvanized scarp can be heated in direct smelting process or electric arc furnace to produce steel or recovery of non ferrous metals. Substantial reduction in CO_2 emissions is expected as the blast furnace use is avoided. Figure 3 shows reduction in CO_2 emissions per ton of steel produced for three different technologies using conventional blast furnace, advanced blast furnace and scrap from IEA study.

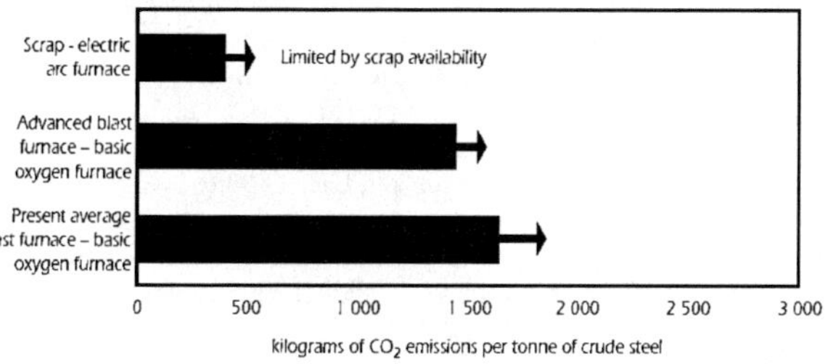

Figure 3. Reduction in CO2 emissions per ton of steel produced (ref [8])

In India, in the total production of finished steel of 36.19 Mt in 2004, production from secondary steel sector was 20.5 Mt. Part of it was from utilization of scrap. Availability of scrap in view of low per capita consumption is the main constraint. There is a need to make more organized effort in this direction.

IV. Need for Waste Heat Utilization

Heat losses in a steel plant are in the range of 17 to 36%. Most emissions from steel industry are related to quenching, gas recovery, casting & rolling and off-gases from different types of furnaces. Only a part of the actual emissions are recoverable. It is estimated that a waste heat utilization programme from hot gas emanations during steel manufacture can save up to 25% of the energy consumed. With this in view, India has proposed several CDM projects in anticipation for Certified Emission Reduction (CER) credits from the technology developed for waste heat recovery and energy efficiency improvements. Another option which needs to be pursued for CO_2 removal from blast furnace top gases is recycling by chemical absorption methods.

V. Development of Non Recovery Type Coke Oven

World wide non-recovery coke making technology has staged a come back with the intrinsic capacity to meet emission control norms and recover waste heat to generate power. By-product recovery coke making oven that produces both coke and by-products such as tar, pitch, benzyl, and ammonium sulphate has been the dominant coke making technology so far. However, reduced value for by-products and competition from the petrochemicals as well as environment concerns have led to adoption of non recovery coke oven as a better option.

Such coke plants are proving less capital intensive and offer higher flexibility in the use of a wider range of coal quality, including semi soft coals. India has demonstrated new standards in steel making by using non recovery coke making technology at SESA coke plant at Amona, Goa [9]. The volatiles evolved during coal carbonization, are not recovered as by-products but are combusted completely in presence of controlled quantity of air and the utilizing heat of the volatiles of evolving gases for coking the coal mass into coke and thus no external heating is required.. It is called non-recovery as there is no recovery of by-products, but it is an energy-recovery coke making process. The SESA coke ovens for the blast furnace are designed with an annual capacity of 280,000 TPA. Process automation could help in reducing capital and operational costs compliant with total emission control and even add to further possibility of using waste heat in power generation

VI. Significance of Carbon Capture and Storage

The average CO_2 emissions from steel industry in India are 2.7 ton / t cs. World average is 1.6 tons / t cs [5]. In order to achieve more stringent CO_2 emission control, the carbon capture and storage (CCS) technology is emerging as means of effective mitigation option for stabilizing CO_2 concentrations. The excess CO_2 is separated from its point sources and converted into liquid form to transport it to desirable locations. For storage CO_2 can be injected in to depleted oil and gas reservoirs. The other possibilities are rocks which slowly react to mineralization, deep saline formations - underground or under the sea bed [10]. Maximum CO_2 in steel making is produced from the blast furnace. The hot gases produced are used for supplying heat in many other processes and CO_2 capture is highly uneconomical. Moreover, there is risk to environment safety from its underground storage. Algae pond with high productivity appears a cost-effective solution. In a steel plant at Vishakhpatanam the flue gas emitted from the blast furnace chimneys is being experimentally tested for CO_2 sequestration using fresh water micro-algae route [11].

Oxygen Blast Furnace

In steel making another option to capture CO_2 from industrial point source is by re-designing the blast furnace for oxygen use. From Oxy fuel plant CO_2 stream in the flue gas is highly concentrated making it easier to capture. Oxy fuel combustion has become the focus of high technology activity worldwide in past two years. However, results from blast furnace are not known. Its efficacy need to be tested in view of the fact that the energy need for an electric arc furnace would be less than 50% of the energy need of an oxygen fuel process in which blast furnace uses most of the energy.

Using Solid Waste

Utilization of steel generated slag is another means of CO2 sequestration. Metal slag is generated in large quantity (up to 200kg per ton of steel) and is highly variable in composition depending on raw material quality. In order to utilize the slag, two alkaline earth metal oxides calcium oxide and magnesium oxide, if in sufficient concentrations, both react with CO_2 in 1:1 molar combination to form Ca CO_3 and Mg CO_3 and they can economically sequester CO_2. In this manner CO_2 sequestration can have benefits in terms of chemical stabilization of the slag mineral, which is utilized in cement production. It has been estimated that using average composition value for CaO and MgO, the slag produced / ton of steel has potential of sequestering 5-10% of BF- Basic Oxygen furnace CO_2 emissions and 28-39% of Electric Arc furnace CO_2 emissions [12, 13]. On average 0.25kg CO_2 per kg of slag can be sequestered. Efforts need to be mounted to increase it up to 50%.

VII. Initiation of Policies and Regulations

a) Worldwide new initiatives have been launched to reduce CO_2 emissions from steel production. After reaching optimum CO_2 emissions, International Iron and Steel Institute (IISI) has started a major R&D initiative on Ultra Low CO_2 Steelmaking (ULCOS); a consortium of 48 European Cos. from 15 countries aimed at development of breakthrough technologies to reduce the CO_2 emission by 50%. It is looking for coke less iron making techniques in BF. Advanced direct reduction processes and conversion of iron ore to metal through electrolysis have been suggested.

b) Critical policy initiatives launched in India from time to time are; liberalization of industrial policy, fiscal incentives, simulation packages and support to R&D. Noteworthy R&D contributions has been made by the industry in taking technology improvements. In 2006, India joined Asia Pacific Partnership in Clean Development and Climate (APP6) comprising of six nations representing over 50% of steel production to address energy issues vis-à-vis CO_2 emissions reduction through technology and process know how transfer.

c) Clean Development Mechanism has been introduced under Kyoto Protocol of UNFCCC for trading of CERs among developed and developing countries. Overall market for CDM projects in India has been estimated to be quite large in the range of 50 to 500 Mt of CO_2 per annum in 2020.

d) Under the National Action Plan on Climate Change, a mission for Enhanced energy efficiency has been proposed. A scheme on Perform, Achieve and Trade (PAT) has been introduced as a market based mechanism. While, for industry PAT provides a vehicle for trading energy saving certificates as a measure of energy efficiency improvements, there is need for developing technology innovation networks as well.

e) Increase in investment for research and developing partnerships for open access to non proprietary information would help in accelerating growth by adoption of new technology as quoted by World Bank study, 'Open access to non-proprietary information agreed at the initial stages of development will be particularly helpful in accelerating research phase of new technology development' [14]. A small beginning in this direction has been made through inter-sectoral collaboration in steel R&D in India.

Conclusions

India has achieved perfection in wrought iron making quite early in 400 B.C. The modern steel industry started after India attained independence in 1947. It has gone through different phases of growth and is aiming to reach world standards. Further advancements in technology can lead not only to lowering of CO_2 emissions, but also production of high strength steel. Steel with higher and higher tensile strength would continue to remain in demand for automobiles and energy industries. The critical inputs would be required for achieving CO_2 reduction by technology management in following areas.

1. Reduction of specific energy consumption by technology up-gradation in different segments.
2. Improving access to clean coal by identification of resources amenable for steel industry
3. Incentive oriented organization of steel scrap for recycling.
4. Technology development for utilization of waste heat
5. Partnership in technology breakthroughs in steel making processes
6. Better utilization of steel slag into products. As such CO_2 capture from hot BF gases is not very effective because steel making process up to finished product is highly integrated.

7. Increase in investments in research, development and demonstration (RD&D). Research collaborations with open access to non-proprietary information should also be encouraged.

Acknowledgements

Author is thankful to Director General, CSIR for the support. Thanks are also due to Dr. Neale R. Neelameggham, Chair – Energy Committee, Co-organizer- Magnesium Committee, TMS2010 for the encouragement.

References

1. European Steel Industry Reaffirms Commitment to Reducing CO2 Emissions, IISI, http://www.allbusiness.com/primary-metal-manufacturing/iron-steel-mills-ferroalloy/414991-1.html
2. Malti Goel, "The importance of R&D and Market Based Mechanism for Technology Management in Steel Industry", *JPC Bulletin on Iron & Steel*, Vol. VI, February 2006, 7-14.
3. http://jpcindiansteel.nic.in/techolsepp.asp.
4. K. K. Singhal, "Energy Efficiency in Steel Industry and Clean Development Mechanism", *Int. Conf. on Clean, Green & Sustainable Technologies for Iron Steel Making* – Bhubneshwar, 2008.
5. "Climate Change & Steel Sector", *Tata Steel in IPCC Working Group III Dissemination Workshop*, Kolkata, October 11-12, 2007.
6. "Pre-combustion Clean Coal Technology Assessment", *TIFAC-STAC study, (TIFAC: RCEIS): 002*, Department of Science and Technology, October 1995.
7. S. K. Gupta and S. S. Gupta, "Corex Process - One of the dynamic routes for gel making with special reference to the success of JVSL"
8. "CO_2 Capture and Storage, A Key Carbon Abatement Option", *IEA Energy Technology* Analysis, ISBN 978-92-64-041400, 2008, 266 pages, IEA Paris
9. HK Chati, GP Kamat, HPUK Nair & PFX D'Lima, "Sesa Energy Recovery Cokemaking: Coke Technology For The Future", *http://www.sesagoa.com/Pdf/SesaEnergyRecoveryCokemaking.pdf*
10. Malti Goel, "Climate Change, Carbon Capture and Storage Technologies and Carbon Trading", *Carbon Capture, Sequestration & Trading* eds B. Das, S. K. Ghosh, S. K. Banerjee, H. S. Roy and S. P. Mukharjee, (Ashirvad Agency, Ghoshpara, Howarah, India, 2009) 1-20.
11. Uma Devi, Private Communication with author, Andhra University, Vishakhapatanam, 2009.
12. Danielle Bonenfant et al, "Molecular analysis of carbon dioxide adsorption processes on steel slag oxides", *Int. J. Greenhouse Gas Emissions*, 2008.
13. S. Bhattachary, S. Mitra Mazumdar & P. K. Tripathi, "Carbon Sequestration in Steel Industry through Waste Utilization", *Carbon Capture, Sequestration & Trading* eds. B. Das, S. K. Ghosh, S. K. Banerjee, H. S. Roy and S. P. Mukharjee, (Ashirvad Agency, Ghoshpara, Howarah, India, 2009) 148-159.
14. P. Avato and J. Coony, "Accelerating Clean Energy Technology Research, Development and Deployment: Lessons from Non Energy Sectors", *World Bank Working Paper No. 138,* 2008.

Energy Technology 2010
Edited by: Neale R. Neelameggham, Ramana G. Reddy, Cynthia K. Belt, Ann M. Hagni and Subodh Das
TMS (The Minerals, Metals & Materials Society), 2010

Development of reverberatory furnace for copper scrap smelting using reformed natural gas

A.KAWY, MOHAMED AHMED[1].
[1]Researcher Assistant, Central Metallurgical R&D Institute.
Keywords: Copper, Reformed, Smelting, Energy

Abstract

Smelting of scrap copper and residual anodes and cathodes carried out in a reverberatory type furnace. In this, deoxidation process is carried out by wood, which acts as a resource of reduction gases such carbon monoxide, hydrogen and methane. Use of wood causes the formation of soot pollutants and efficiency of oxygen removal from copper is low leading to added costs in removing contaminants. This expensive problem was solved by using of reformed natural gas as a source of reduction gases ($CO+ H_2$).

Introduction

Three reverberatory furnaces were used for smelting of scrap copper and residual anodes and cathodes. Each furnace had a 7tpd capacity and was operated at 1183 C.
In this furnace, Step1-compressed air supplied by tubes from the side of the furnace to remove sulfur from copper matte by oxidation .The copper is also oxidized to copper oxide .[1]
Step 2- copper from matte is deoxidized by a reductant. Typically deoxidation process is carried out using wood as a resource of carbon monoxide, hydrogen and methane which are the reduction gases. Use of wood causes the following:
1- Forms soot pollutants
2- Oxygen removal efficiency from copper is low; more contamination in copper matte increased energy in deoxidation
3-the process is highly expensive.
In the recent trends, this problem has been solved by using reformed natural gas as a source of reduction gases (CO+ H) .This results in
1-Reduced pollution during smelting.
2- High efficiency in the removal of oxygen from copper to produce oxygen free copper. The surface of copper produced from this method is smooth which helps in forming and machining.
3- Low cost and natural gas is more readily available in Egypt than wood.
This modification to the use of reformed natural gas is done without changing the design of the furnace.

Reformed gases are introduced into furnace as the air is introduced in stage 1. The gas recycling from reduction process to the reformer is done by closing the furnace after copper scrap is fed.

This is the main objective of the development of rotating furnace with the use of reformed natural gas.

Experimental work

Process description

Air is compressed using a Root's blower to about 1.4 bar (abs). This ratio between combustion gas and air is kept constant by means of automatic ratio control. In addition, the flue gas is analyzed continuously using a H_2 –analyzer. In case of deviation of H_2-content, the ratio is immediately corrected. Thus the H_2 content is kept constant. [5]

The combustion is carried at the pressure of 300 mbar (g) in a refractory lined reaction chamber with sub-stoichlometric air according to the equation

$$C_nH_m + O_2 + N_2 \longrightarrow CO_2 + H_2O + CO + H_2 + N_2 \qquad (1)$$

The reaction chamber is equipped with an electrical ignition and with an ignition burner as well as an automatic self –controlling UV-flame monitoring device (see figure. 1.1). The hot flue gas generated by this process is fed to a boiler for Mono-ethanol amine regeneration by the pressure –less regeneration-strip-steam.
The CO_2,is scrubbed by (see figure .1.2) a regenerative organic alkaline solution (MEA) in an aqueous medium with 12.5 wt% as CO_2-absorbent in the two CO_2-scrubbing towers, the reaction takes place at about 25°C according to the equilibrium:

$$CO_2 + H_2O + NH_3CH_2CH_2OH \rightleftarrows (HOCH_2CH_2NH_3^+)\ 2CO_3^- \qquad (2)$$

The saturated and hot alkaline solution goes to the CO_2–stripper tower where the CO_2 is removed via strip stream generated in the MEA boiler.
The CO enriched gas is cooled and the condensate is routed back to the MEA boiler, the HNX –gas coming from the CO_2-scrubbers is pre dried via a heat exchanger. The temperature of the cooling water should be maximum 8° C, the subsequent final drying process is affected via a thermally regenerated double adsorption plant. As adsorbents molecular sieves are used.
While dehydration takes place in the first vessel, the second vessel is regenerated via a circular blower at temperatures exceeding 200° C. the

heating –up process to regeneration temperature is affected via a gas heat exchanger, then, the compression of CHN gas to 26 bars (abs) is done by means of three stage dry running, reciprocating compressors.
CHN gas coming from the gas compressors can be stored in gas storage tanks (see figure 1.3)
CHN –gas injection into the revabratory furnace through four double – walled tuyers located along a line in the side of revabratory furnace (see figure.2)
The tuyers operate in one four possible modes:
1-repose-when no metal is covering the tuyers
2-mixing –to simply keep the melt stirred
3-reduction
4-oxidation
(See figure. 1)

Technical Data: All technical data refer to one single plant and the capacity and composition of CHN gas in the following table (see table I):

Capacity ± 5%	Composition					
	H_2 % by vol.	CO % by vol.	CO_2 VPM	O_2 ppm	CH_4 VPM	N_2&Ar balance
1000 Nm^3/h	0.5-15.2	10.5	100	5	100	

* -

Dew point: -60°Cat 1031mbar, 0° C
-Delivery pressure in bar (abs):
Exit CHN-plant: 0.05 bar (g)
Downstream the compressor: 25 bar (g)
-Dimensions: approx. 4m× 20m× 10m height
-Total weight: approx 70.000 kg
Consumption Data: Natural gas at 15 % by vol., % H2: approx. 170 Nm^3/H.
-Electrical energy :(3 × 400 V, 50Hz), voltage± 5%, frequency: ± 2%
Consumption:
CHN –plant: approx: 102 KW
-CHN-compression, 25 bar: approx. 202 KW
Total: 304 KW
-Cooling water :(supply pressure = 4 bar)
CHN-plant:
-t_{in} =25°C, dT=18° C: APPROX. 73 m^3/h
Compression CHN-Gas, 25 bar (g):
-t_{in}=25° C, d T=10° C: approx. 16 m^3/h
Instrument air :(supply pressure p= 6 bar, dew point=-40° C at 1013 mbar, 0° C)

-instrument air: approx. 16 Nm^3/h

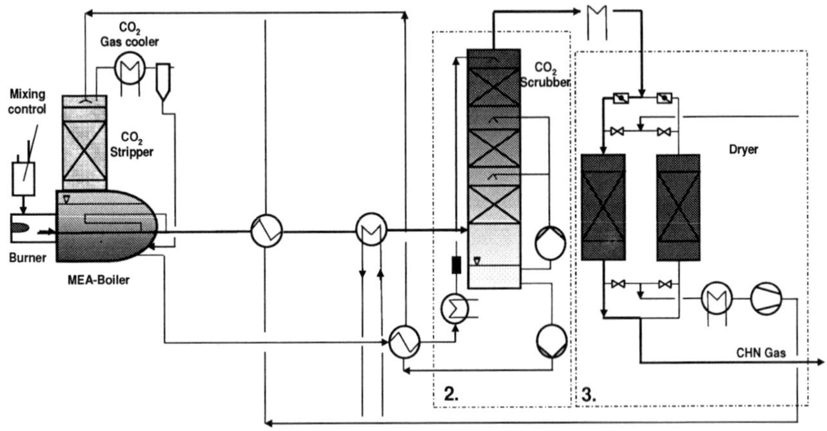

Figure.1, scheme of reformed natural gas plant

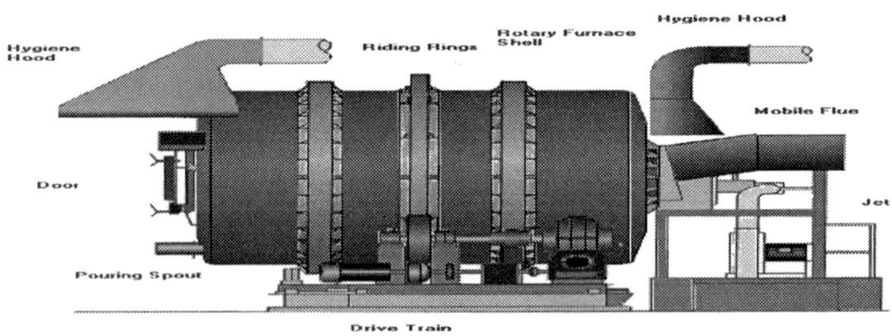

Figure.2, developer reverberatory furnace will use in ECW before connecting of CHN –gas plant

Result and discussion

Solubility of hydrogen in liquid copper in the furnace is a very important factor to control. The main equation for this state is

$$H_2 (g) + [O] \longrightarrow H_2O (g) \qquad (3)$$

Hydrogen is soluble to a limited extent in liquid copper, and its equilibrium concentration is given by: [4]

$$K = \frac{\rho H_2O\ (atom)}{[wt.\ \%H]^2\,[wt.\ \%O]}$$

Which has values of 5×10^9 at 1150°C and 15×10^9at i083°C(Allen and Hewitt, 1933).Thus, the final hydrogen content to liquid copper is in the order of 2×10^{-5}wt.%

Hydrogen and Oxygen decrease in solubility during cooling and solidification, and the combine to form steam by the reaction:

$$2[H] + [O] \rightleftharpoons H_2O\ (g),\ 1\ atom \qquad (4)$$

The equilibrium solubility of hydrogen in solid copper (Cu2 O saturation) is given by the expression:

$$K = \frac{\rho H_2O}{[wt.\ \%H]^2}$$

Where K (1000 °C) is 10^{12}. At 1000 °C, ρH_2O= 1 atm (evolution of pure steam) the equilibrium hydrogen content of solidified copper is, therefore, 10^{-6} wt. %.[2]

The volume of steam evolved during cooling and solidification can be calculated from change in hydrogen concentration during these steps, using the equation (2). The volume (1000°C) of steam evolved in proceeding from the conditions after reduction:

$[H] = 2\times10^{-5}$ wt. %, [O] = 0.1 wt. %

To the conditions after solidification:

$[H] = 1\times10^{-6}$ wt.% , [O] = 0.005 wt.% , Cu_2O saturation

Is approximately 0.1 cm^3 of H_2O per cm^3 of solid copper? This is somewhat larger than the volume of steam required to compensate for metal shrinkage (0.05 cm^3/ cm^3 of solid copper) but some steam will escape from the metal during the solidification process.

The actual amount of steam evolved depends upon the hydrogen content of molten metal. This increases with the degree of oxygen elimination during deoxidizing process, and cooling rate. it appears, however , that an oxygen content of 0.05-0.2 % after deoxidation and its equivalent hydrogen content (~ 2×10^{-5} %) , provide conditions which are suitable for obtaining “flat set ” copper .[3]

Conclusion

Refining of copper scrap by reformed natural gas offer numerous benefits, including:

1-Premium quality refined copper from lower cost charge materials

2-higher energy efficiency

3-Qualty levels (low dissolved oxygen) previously unviable to fire refining

4-Higher productivity from modifying existing equipment

5-Reduced pollution and environmental concerns

6-High productivity per labor man hour.

Acknowledgment

The author is grateful to Josef latezko, Mahler company for help in this project.

References

1-A.A. Tsidler, *Metallurgy of copper and nickel*, (Jerusalem, Israel program for scientific translations, 1964), 140-141
2-Foseco, *Non ferrous foundaryman`s handbook* (New York, NY: pergamon press, 2000), 240-245
3-Ffernando Gomez, June Ignacio Guzman, john e. tilton, "*Copper recycling and scrap availability*" (Elsevier company, 2007)
4-Tilton, i.e., "*The future of recycling*", (policy24 (3), 1999), 197-204
5-Valencia, c., *An econometric study of the world copper industry* (Ph.D.Disseration, Colorado school of mines, 2005)

Energy Technology 2010
Edited by: Neale R. Neelameggham, Ramana G. Reddy, Cynthia K. Belt, Ann M. Hagni and Subodh Das
TMS (The Minerals, Metals & Materials Society), 2010

BAUXITE RESIDUE NEUTRALIZATION WITH CARBON SEQUESTRATION

Luis Alves Venancio[1], Antonio Ernandes M Paiva[1], Emanuel Negrão Macedo[2], José Antonio Silva Souza[2]

[1]Instituto Federal de Educação, Ciência e Tecnologia do Maranhão, PPGEM
Av. Getúlio Vargas,04, Monte Castelo, São Luis, Maranhão, CEP 65025-001, Brazil
[2]Universidade Federal do Pará - PRODERNA
Rua Augusto Correia 01, Guamá, Belém, Pará, CEP 66075-110, Brazil

Keywords: CO_2, Carbon, Residues, Recycling, Bauxite, Alumina, Global warming.

1 Abstract

The production of alumina from bauxite using the Bayer process generate 1.25 ton of the residue known by red mud and 460 kg of CO2 per ton of alumina.
The direct use of exhaust gases from the alumina refineries to react and neutralize the bauxite residue may allow a double gain: open a wide range of new applications for bauxite residue reducing the reactivity of red mud and sequester approximately 15% or 67 kg of CO2 per ton of alumina.
This paper shows the reaction of a suspension of bauxite residue-water with a simulation of the calcinator exhaust gases of the refinery. The influence of the temperature and the ratio water red mud is also analyzed.

2 Introduction

The production of millions of tons per year of alumina around the world generates around 1.25 tones of the residue known as red mud and 0.4 ton of CO_2 per ton of alumina . The vast majority of the red mud is stored in large sealed lakes. Due to the high reactivity is considered dangerous and currently there is no operating process capable of reuse all the material at any refinery. These lakes after filled of solid material eventually have the surface revegetated. Otherwise will need monitoring for decades and remain as an environmental passive. The search of a industrial process that combine the reduction of the CO_2 emission and the reduction the pH of the red mud opening the possibility of use in large scale in a economically viable way is the main objective of this work.
The use of the exhaust gases of the refineries to neutralize the bauxite residue may allow a double gain: Open a large array of new applications for the residue with the reduction of its reactivity and sequester million of tones of CO_2 plus other gases like SO_2 [7] and NO_x.
This project main objective is to study the reaction of the CO_2 with the Bayer process residue in order to obtain the basic parameters to design an equipment capable of process this reaction on a pilot scale. With this objective compositions of red mud on spray were processed with flow of gas containing CO_2 at different temperatures and with concentration similar to the exhaust gases of the calcination process of the refineries. With the results, basic concepts of a pilot scale equipment were evaluated.
A revision of the work already done on this type of residue were done [3, 9, 10]. Alternatives and potential barriers on chemistry and environmental fields were discussed. With the objective of industrial application the potential use of these materials were evaluated through the pH control, mass balance and x-ray diffraction.

3 Process Review.

3.1 Bayer Process For Alumina Production

The Bayer process to obtain alumina from bauxite used since 1893 [5] can be divided in four steps:

3.1.1 Digestion – The minerals containing aluminum at the bauxite are selectively extracted from the non soluble components through the dissolution in sodium hydroxide. Different bauxites have various mineral contents.

Gibbsite:

$$Al(OH)_3 + Na^+ + OH^- \Rightarrow Al(OH)_4^- + Na^+ \quad (1)$$

Boehmite and Diaspore:

$$AlO(OH) + Na^+ + OH^- + H2O \Rightarrow Al(OH)_4^- + Na^+ \quad (2)$$

3.1.2 Clarification – After the extraction the insoluble residue must be separated through a settling and filtering process. The aluminum containing liquor is transferred to the precipitators.

3.1.3 Precipitation – Crystalline aluminum trihydroxide "hydrate" is precipitated from the liquor.

$$Al(OH)4^- + Na^+ \Rightarrow Al(OH)3 + Na^+ + OH^- \quad (3)$$

The hydrate is fed into the calcination kiln.

3.1.4 Calcination – The hydrate is calcined to form alumina trough the separation of the water.

$$2Al(OH)3 \Rightarrow Al2O3 + 3H2O \quad (4)$$

3.2 Bayer Process Residue – Red Mud

Alumina production through Bayer process generates from 0,7 to 1,3 tons of residue per ton of alumina. It is usually called red mud (RM). The RM composition include hematite and e other iron oxides, quartz and titanium oxides.

On the RM treatment only part of the sodium hydroxide is removed. The RM have a considerable amount of sodium hydroxide and several concentrated ores. This represent a considerable resource waste of mineral resources. Table 1 shows the chemical analysis of the RM from Alumina do Norte do Brasil S/A (ALUNORTE – Belém-PA-Brazil).

Table I. Dry Red Mud Composition

Content	(%)
Fe2O3	34
TiO2	5
CaO	1,3
SiO2	18,3
Al2O3	22
Na2O	9
K2O	0,12
MgO	0,04
V,Ga, P, Mn, Mg Zn, Zr, Th, Cr, Nb	Traces

3.3 Residues From the Bayer Process – Exhaust Gases

The generation of exhaust gases at the Bayer process occurs essentially at two points: the steam pgenerating boiler and the alumina calcinatory. The calcinators usually use fuel oil or natural gas.
The specific consumption is 0.18 ton of oil for each ton of alumina [1].
The composition of the exhaust gases can be estimated considering the oil have 80% carbon on its composition and reacting with oxygen each atom of carbon with atomic weight 12 will meet two of oxygen with atomic weight 16.

$$C+O_2 \Rightarrow CO_2 \quad (5)$$

Therefore each ton of oil will generate 2.9 tons of CO_2

3.4 Red Mud Carbonation

The RM has around 5% of NaOH [1] on its composition.
In presence of H_2O as a catalyzer:

$$NaOH + CO_2 \Rightarrow NaHCO_3 \quad (6)$$
$$(22{,}99+16{,}00+1{,}01= 40) + (44{,}01)$$

To neutralize 1.3 ton x 5% NaOH of RM for each ton of alumina, we need 71.5kg of CO_2. Since we have a minimum of 396 kg of CO2 for each ton of alumina, we have excess gas for the neutralization.

4 Equipment

In order to study the reaction of CO_2 with the RM we processed compositions of RM and water with gas streams containing CO_2 concentration is similar to the exhaust gas of calcination at different temperatures.

The experimental system used in this study whose schematic is shown in Figure 3.1, consists of a washer type spray tower, with cylindrical geometry that operates vertically, a reservoir, a pump, a fan, valves and measurement equipment. The suspension is transported from the reservoir by the pump to the sprinkler system inside the washer. The injection of the suspension is made by 10 hollow cone nozzles, stainless steel 304 and the opening of 2 mm diameter or by two or four nozzles centrifugal hollow cone sprays.
The scrubber, where the experiment was performed, has the main objective of promoting the reaction between carbon dioxide (CO_2) from the gas stream with caustic soda (NaOH) present in the liquid-solid phase. The gas stream of air mixed with CO_2 is fed by direct entry of gas at the bottom of the washer, which comes into contact in countercurrent with the recirculating suspension of RM.

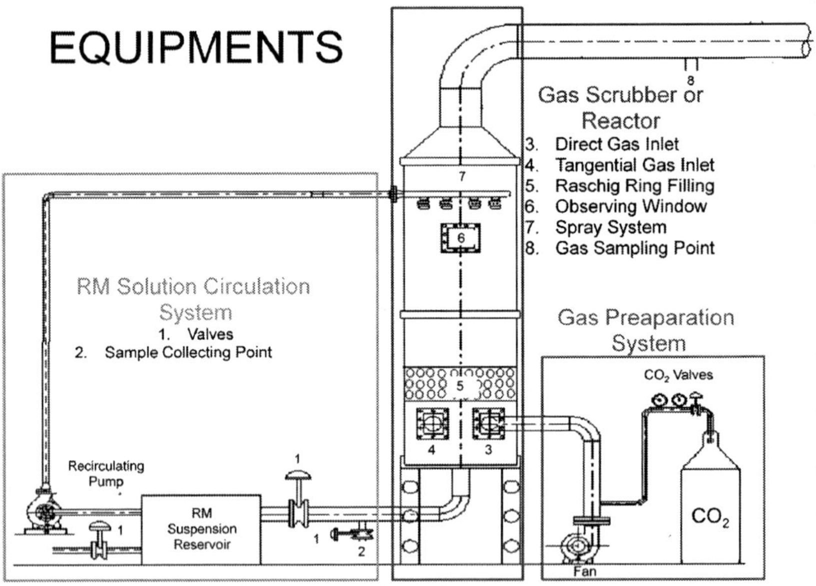

Figure 1 – Experimental gas RM reactor

4.1 Gas Supply Preparation System

It consists of a ventilator that blow air through a hose of 100 mm in diameter to the direct input of the washer. At one point downstream of the blower bottled CO_2 is injected by a flexible hose which is attached a diffuser, allowing a homogeneous mixture of the gases before entering the washer and prevents clogging of the CO_2 circuit by the formation of ice.

4.2 Scrubber or Reactor

The washers of the type spray tower main function is to provide contact between the two phases (liquid and gas), thereby occurring the transfer of CO_2 from the gas mixture to the liquid absorber.

The scrubber used in this experiment was designed according to [2, 4 , 8] and is mounted at the Laboratory of Chemical Engineering at UFPA.
The scrubber consists of a cylindrical shell of glass fiber with 10 mm thick. It has 2 meters height, 60 centimeters in diameter and is divided into two symmetrical parts (1m), joined by a sealed flange set screws with 10 stainless steel 3 / 8"x 1"and a rubber gasket of 1 / 4"thick.
At the base of the washer there are two inlet for the gases, one is a direct input, or input that directs the gas through the center of the washer and the other directs the gas through the walls. In this experiment we used only the direct input. Above the entries there is a system of filling type Raschig rings, consisting of PVC pipes of 20 mm in diameter with various sizes, with the aim of increasing contact area of the gas suspension washing. The washer has at its base than a PVC pipe of 40 mm which serves as the outlet for suspension washing.
Near the top is located a viewing window made of glass with 570 cm2 area that allows the observation of the sprinkler system. This consists of 10 nozzles (hollow cone) or by two or four nozzles industrial hollow cone type centrifuge, distributed in a radial washer providing drop distribution in cross section as homogeneous as possible and a top pipe used to output of gases to the atmosphere. Gases contaminated following an upward trend in counter-current with drops of suspension, provided by the liquid absorber.

Figure 2 – View of the Scrubber

4.3 RM Solution Circulation System

The liquid absorber was pumped from the recirculation tank. The pH was measured by digital

equipment that allows monitoring throughout the experiment in order to identify the saturation point of the suspension.
At this tank was mounted an electric heating system with 4kw of power that raised the temperature of the suspension of mud to 48 ° C in the experiment with heating.

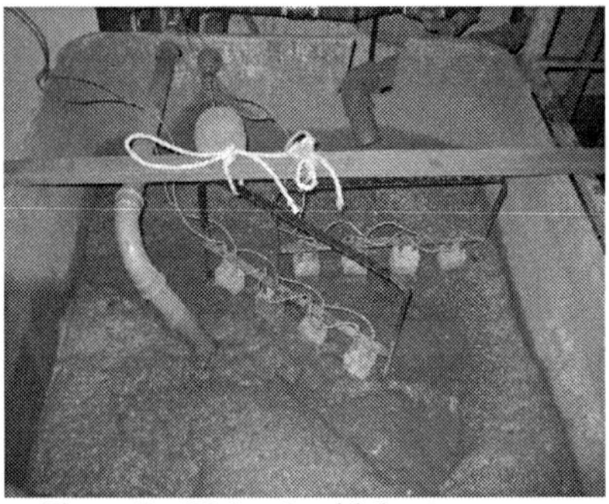

Figure 3 – RM tank with heater

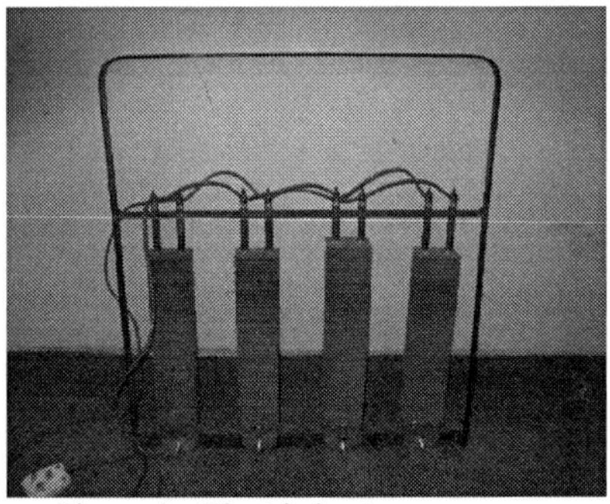

Figure 4 – Electric heater

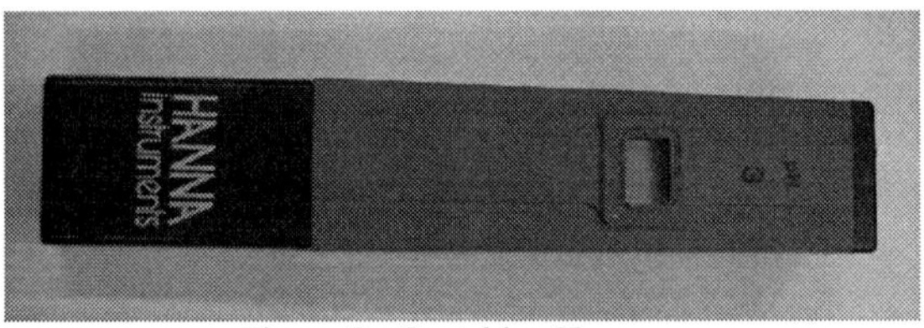

Figure 5 – Portable pH gauge

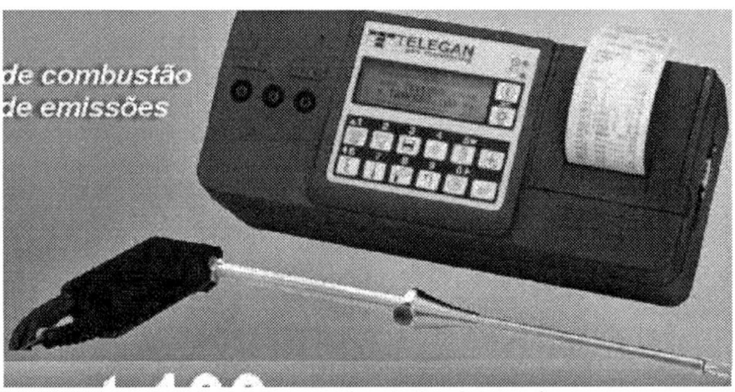

Figure 6 – Gas analyzer

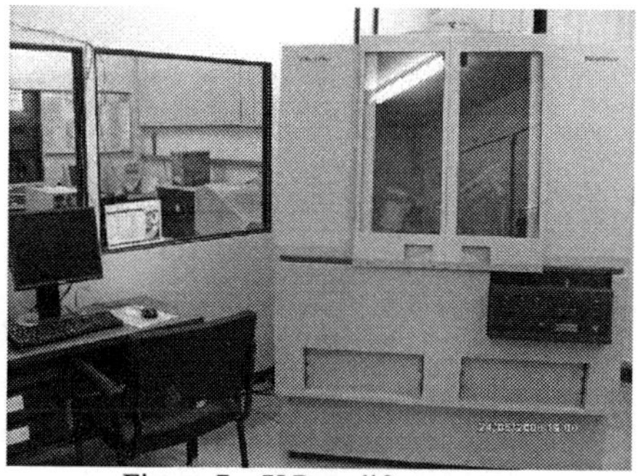

Figure 7 – X Ray difratometer

5 Analysis of results and interpretation of data.

The goal was to work with a content of 10% CO2 gas to simulate the effluent gas from a real calciner. This gas was able to reduce the pH up to three points.
We tested two types of nozzles. There were no significant differences in the results.
The increase in temperature at 20 ° C caused a 50% reduction in reaction time as we can see in the graphs below pH x Minutes ..

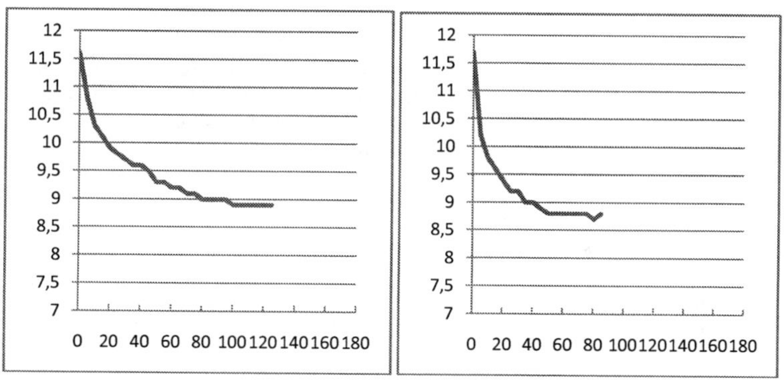

Figure 8. On the left the temperature was 28 °C at the right 48 °C

The sample VE-01 is composed of the mineral phases Natrita (Na2CO3), red in the diffractogram, the best fit was obtained with the form 37-0451 (ICDD-PDF), and the phase Dawsonite (NaAlCO3 (OH) 2), blue in the diffractogram, item 45-1359.

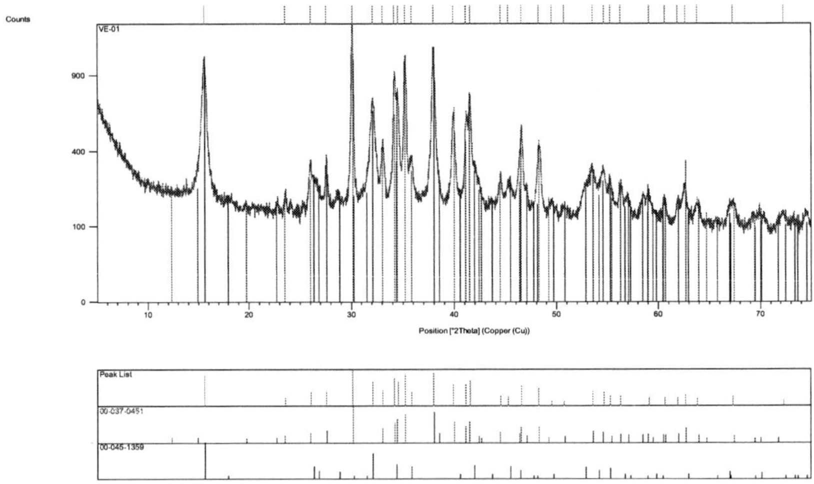

Figure 9. The result of X-ray diffraction confirms the formation of the carbon compounds.

6 Conclusion.

The direct use of gas effluents from the calcinator to reduce the pH of red mud and carbon sequestration is a viable low cost solution of reacting two effluents from the same process to reduce the environmental impact. The next step is to build a prototype in a pilot scale to confirm and optimize the rates of mass transfer. Carbon compounds formed are water soluble and therefore leachable. Carbonation does not make the product ready to be disposed in nature but makes it much less reactive which will certainly open a whole new range of possible applications in large scale. This is another field to be searched.

7 References

1. ALUMAR e ALUNORTE – Alumina refineries sites located at Pará and Maranhão states, Brazil. http://www.alumar.com.br e http://www.alunorte.net

2. CALDAS e LACERDA, Torres Recheadas. Coleção Tecnologia Brasileira. JR Editora Ltda. Rio de Janeiro. 1988.

3. COOLING David; SMITH Peter. Ecos magazine, November 2005. http://www.publish.csiro.au/?act=view_file&file_id=EC127p35.pdf

4. HESKETH, H.E. Air pollution control. Pennsylvania, Technomic Publish, 1996.

5. IAI - International Aluminum Institute – http://www.world-aluminium.org - The Institute has 25 Member companies and represented more than 80% of world primary aluminum production.

6. LME London Metals Exchange – http://www.lme.co.uk .

7. LOBATO, Ivan da Costa. Study of a gas scrubber to control SO_2 emissions. 1998. 137f. Dissertation (Master in Chemistry Engineering) – Universidade Federal do Pará. Belém, Brazil, 1998.

8. PERRY e CHILTON (1980) Chemistry Engineering Handbook, 5a Ed.. Rio de Janeiro: Guanabara Dois, 1980.

9. SHUNMEI Shi (2001) Carbon dioxide sequestration via pH reduction of red mud using liquid CO_2

10. GUILFOYLE, L*, Hay, P, Cooling, D. Use of flue gas for carbonation of bauxite residue. 2005. Proceedings of the 7th International Alumina Quality Workshop.

Energy Technology 2010
Edited by: Neale R. Neelameggham, Ramana G. Reddy, Cynthia K. Belt, Ann M. Hagni and Subodh Das
TMS (The Minerals, Metals & Materials Society), 2010

Voltammetric measurements for CO_2 reduction under ambient conditions

Maria D. Salazar-Villalpando,* Bryan Reyes and Anthony Cugini
National Energy Technology Laboratory, US Department of Energy, Morgantown, WV 26507, USA

*Corresponding author. Tel.: 01 304 285 5427; fax: 01 304 285 4469.
E-mail address: maria.salazar@netl.doe.gov (M.D. Salazar-Villalpando)

Key words: CO_2 reduction, Voltammetric measurements, $NaHCO_3$, and $KHCO_3$

ABSTRACT

The electrochemical reduction of CO_2 was performed in a laboratory-made, divided H-type cell. A copper electrode was used as the cathode. An Aldrich Nafion 117-type ion exchange membrane (0.18mm thickness) was used as the diaphragm. $NaHCO_3$ and $KHCO_3$ were studied as the catholytes. Hydrogen evolution from water reduction and products from CO_2 electrochemical conversion were recorded. Higher CO_2 reduction enhancement was observed by $KHCO_3$ as compared to $NaHCO_3$. The main conclusion of this work is that in fact CO_2 can be electrochemically reduced with copper electrodes.

1.-INTRODUCTION

The direct electrochemical reduction of CO_2 may allow for a simple process and, by avoiding high temperature reactors, would enable the process production rate to be quickly varied to follow the time of day availability of surplus electricity [1]. Temperature has a major effect on the electrochemical conversion of CO_2 since its solubility increases when temperature decreases [2]. The electrical potential applied to the electrochemical reacting system will have an influence on the product distribution, as well. For example, a narrow potential range during cyclo-voltagraphic studies will determine the detection of a lower number of products. A recommended potential range to study the electrochemical reduction of CO_2 is -2V to 2V. Moreover, the reaction rate on the electrode is governed by various processes, such as the kinetics of electron transfer, adsorption/desorption at the electrode surface, and diffusion of CO_2 to the Cu electrode. The availability of active species, CO_2 for electro-reduction, is extremely low at the vicinity of the electrode surface and it seriously interferes with the rate of mass transfer. In batch type electrolytic reactors, depletion of CO_2 at the surface takes place first. CO_2 from the bulk then has to diffuse to the electrode surface, resulting in mass transfer controlled kinetic phenomena. During the above process, several other competitive side reactions could suppress the primary reaction even though the cathode has good electro catalytic properties for reduction of carbon dioxide. When the dynamic supply of specific active species is ensured to electrode surface, the mass transfer effects are minimized to certain extent. Hence it is envisaged that a continuous flow reactor is more efficient for the electro-reduction of CO_2 in CO_2 saturated aqueous electrolytes. The primary reactions that occur at the copper electrode during the reduction of CO_2 are listed below (with the standard potentials calculated using formation energies [1] :

$2H^+ + 2e^- = H_2$ Eo=0.0 V vs. SHE (1)

$2CO_2 + 12H^+ + 12e^- = C_2H_4 + 4H_2O$ Eo=0.079 V vs. SHE (2)

$CO_2 + 8H^+ + 8e^- = CH_4 + 4H_2O$ Eo=0.169 V vs. SHE (3)

$CO_2 + 2H^+ + 2e^- = CO + 4H_2O$ Eo=-0.103 V vs. SHE (4)

$CO_2 + 2H^+ + 2e^- = HCOO^-$ Eo=-0.225 V vs. SHE (5)

Generally, hydrogen evolution is a competing reaction of CO_2 reduction. Therefore, the suppression of hydrogen formation is a challenge. The reaction mechanism during the electro-reduction of CO_2 is still not clear. Kaneco et al. suggests that hydrocarbon formation starts with CO_2 combining with a proton and electron to form HCO_2^-. After this first electronation step, the intermediate undergoes a second protonation/electronation to produce adsorbed CO as the key intermediate. By a succession of four protonation/electronation steps, an adsorbed reactive methylene group forms, and may stabilize as either a methane molecule by a subsequent double protonation/electronation sequence or dimerize to ethylene [2]. In the electrochemical reduction of CO_2, the type of the cathode has a major influence in products selectivity and CO_2 conversion [3]. The behavior of pure metals for this reaction has been fairly well understood. High Faradaic efficiency or selectivity for CO_2 electrochemical conversion is achieved on metals with high overpotential for hydrogen generation, such as Cd, In, Pb, Hg, Ag, or Cu. However, high energy efficiency for the reduction is not achieved [4]. Early studies by Hori et al. have shown methane and ethylene production from the electrochemical reduction of CO_2 over copper electrodes with faradic efficiencies over the range 87-92% [5]. It has been observed that Cu can further convert CO to more reduced species, producing hydrocarbons, partially because Cu seems to have a weak CO adsorption [6]. In metals with a high CO adsorption strength, such as Pt, Ni, Fe and Ti, CO_2 is reduced to form tightly adsorbed CO. Because of the low turnover of the adsorbed CO, if the electrochemical process is driven to 5–10 mA/cm^2, the main product is Hydrogen [7-8]. It has been frequently claimed that poisoning arises from carbonaceous adsorbates on the surface during the electrochemical reduction of CO_2 on copper surfaces [9].

The objective of this work was to conduct voltammetric measurements during the electrochemical reduction of CO_2 and study the effect of the type and concentration of the catholyte.

2.-EXPERIMENTAL

The electrochemical reduction of CO_2 was performed in a laboratory-made, divided H-type cell. An Aldrich Nafion 117-type ion exchange membrane (0.18mm thickness) was used as the diaphragm. This membrane selectively allowed the transport of H^+ ions from the anode chamber to the cathode chamber along with the associated water molecules. The cathode potential was measured with respect to an Ag/AgCl saturated KCl electrode. The volume of the solution was 20 ml. Sodium bicarbonate was used as the ionosphere in the catholyte. Acidic solution (H_2SO_4, 0.1M) was chosen as the anolyte. The counter electrode was a Pt wire. A discontinuous electro-reduction procedure was followed. This is, first, CO_2 gas was bubbled into the catholyte for 1 h. Then, this CO_2-saturated solution was reduced electro-chemically at cathodic polarizations in the range -1.2V to +1.0V versus Ag/AgCl; saturated KCl. Cyclic voltammograms (CVs) at the Cu electrode in the aqueous solution were recorded at ambient temperature and pressure. The sweep rate was 50mV/s. A CV was recorded prior to CO_2 saturation, another CV was recorded right after CO_2 saturation, and then fourteen CVs were continuously recorded to determine steady state conditions. $NaHCO_3$ and $KHCO_3$ were studied as the catholytes.

3.-RESULTS

Effect of electrode polarization

Figure 1 shows the CVs for the 2.0 M $KHCO_3$ aqueous solution, before CO_2 saturation (a), right after CO_2 saturation (b), and the fourteen CV after CO_2 saturation (steady state) (c). Regarding curve (a), it is observed that peaks were detected at -1.2V, -0.88V, -0.63V and -0.08V. According to previous reports, it could be assumed that the first peak at -1.2V, corresponds to Hydrogen evolution from water reduction. The peaks at -0.88V and -0.63V would correspond to copper reduction and the peak at -0.08V to copper oxidation [10]. Regarding graph (b), the new peak at -0.98V would correspond to CO_2 reduction, possibly hydrocarbons. It has been suggested that H_2 is produced on the surface with much Cu oxide, while CO is produced on the surface with predominant Cu metal. CH_4 is produced on the surface with both Cu oxide and metallic Cu, having a high affinity for adsorbed oxygen. C_2H_4 is also produced on the surface with a high activity for CH_4 production. This also suggests that the surface with Cu oxide to some extent prefers hydrocarbon production [11]. In Figure 1, the increase in the size of the peak at +0.08V would correspond to oxidation of CO_2 products and copper oxidation. Curve (c) showed that the peak at -0.98V was not present in this CV, which may be explained by assuming that CO_2 has been completely consumed or that the deactivation of the copper electrode occurred. It has been reported that the deactivation of copper occurs regardless of the extent of its purity. This deactivation has been studied by several researchers and suggested that copper electrode degradation may occur to the deposition of poisoning species such as graphitic carbon, adsorbed organic intermediate or copper oxide. Thus, in our laboratory, further experiments were conducted to determine if in fact CO_2 had been completely consumed or that the deactivation of the copper electrode occurred. The same solution was saturated again with CO_2 for 1 hour and another CV was recorded. Comparison of the first CV ("fresh electrode" and first CO_2 saturation) with the 15th CV (taken after the second CO_2 saturation) showed that the peak at -0.98V increased its size, this is higher current is recorded. In the same way the oxidation peak at positive potential increased its size notably, which confirms that the peak at -0.98V represents the electrochemical reduction of CO_2, Figure 2. In a previous report, studying the electrochemical reduction of CO_2 on copper electrode and FTIR experiments, it was observed that, under Ar atmosphere, CO is adsorbed at the potential where hydrogen evolution occurs. It was also found that adsorbed CO suppresses hydrogen evolution. FTIR experiments in the infrared region, 2300-1700 cm^{-1}, showed absorption bands of adsorbed hydrogen or hydrogen-metal bond of hydrido-carbonyl complex as well as adsorbed CO. Moreover, it has been assumed that adsorbed CO on Cu electrode surface is below the potential of the cathodic charge transfer, and not above the potential. Voltammetric measurements have shown that charge transfer is accompanied with CO adsorption on Cu electrode at -1.0V in a 0.2M $KHCO_3$ aqueous solution [12-14].

Effect of electrolyte concentration

Figure 3 includes the CVs for the 0.6, 1.0, 2.0 and 3M $KHCO_3$ aqueous solution right after CO_2 saturation. It is observed that the size of the peaks at -0.98V and +0.08V increased when the $KHCO_3$ concentration increased. As mentioned before, the peak at -0.98V may correspond to CO_2 reduction and the peak at +0.08V to the oxidation of CO_2 products. According to previous references [15-16], where the electrochemical reduction of CO_2 was conducted on modified Cu electrode (CuCl electrode with KCl as the catholyte), it was suggested that CO_2 may be inserted into the double layer on the electrode in such a way that adsorbed Cl^- and K^+ ions linked with the electrophilic carbon and the nucleophilic oxygen of CO_2, respectively. Similarly, in this report, we suggest that CO_2 links to HCO_3^- and K^+, generating intermediate specie prior to its electro-reduction. This fact would explain that the reduction peak at -0.98V increases as a function of $KHCO_3$ concentration. Figure 4 shows current vs. voltage for this experiment at different $KHCO_3$ concentrations, it is clear that while maintaining a constant potential, the current increases at higher $KHCO_3$ concentration. A previous report has been established that only dissolved CO_2 molecules take part in the reduction process, and not bicarbonate HCO_3^- or carbonate CO_3^{2-} ions from solution, when $NaHCO_3$ or $KHCO_3$ are used as catholytes [17].

The main finding in our experiments is that the peak height of the CVs around -0.98V increases with increasing the concentration of $KHCO_3$ in the presence of CO_2. Cabrera et al. have conducted experiments with CO_2 saturated $KHCO_3$ aqueous solutions with concentrations between 0.01 and 0.5 M at 20 °C. They observed that the lower limit of their CVs was shifted to more positive potentials as the $KHCO_3$ concentration was increased. These results were explained by assuming that this shift could be due to a change in the potential of the quasi-reference electrode (AgQRE) with changes in the $KHCO_3$ concentration. Moreover, they explained that the peak current increase was also due in part to H^+ reduction [18]. It has also been suggested that at high pHs, above 9, adsorbing HCO_3^- may be transformed into non-adsorbing CO_3^{2-}. A reversible adsorption of these species on Pt electrode was reported to occur at pH values above 9. Adsorption competition between Hydrogen and carbonate species has also been suggested [19].

Effect of the type of cation in the electrolyte

$NaHCO_3$ and $KHCO_3$ were chosen as the electrolytes to study the effect of the type of cation in the electrolyte. Figure 5 shows the comparison of these results. It is observed that higher currents were observed in the case of $KHCO_3$ as compared to $NaHCO_3$. Figure 5 shows the Current vs. Voltage, comparison of these results. Higher currents were observed for the former electrolyte, which indicates that it shows a higher promotion for CO_2 electro-reduction than $NaHCO_3$. Previous reports have shown that the type of the catholyte is an important factor in the product selectivity. Electro-reduction of CO_2 on a Cu-foil electrode lead to the formation of C_2H_4, CH_4, and alcohols in KCl, $KClO_4$, and K_2SO_4 solutions of high pH=7 and to the generation of CO, CH_4, and C_2H_4 in $LiHCO_3$, $NaHCO_3$, and $CsHCO_3$ solutions [20]. With a glassy carbon electrode on which Cu was in situ deposited, CO_2 could be reduced to CH_4 and C_2H_4 at 0°C in 0.5 M $KHCO_3$ or 0.25 M K_2SO_4 solution containing $CuSO_4$ [21,22].

4.-CONCLUSIONS

The reduction of CO_2 to hydrocarbon products with copper electrodes is a complex multi-step reaction involving shared intermediates and multiple reaction pathways. The reaction is sensitive to the electrode surface structure, local conditions such as pH and CO_2 concentration (which are in turn influenced by buffer strength, the degree of mass transport, and the bulk CO_2 concentration/pressure) and temperature. Other factors, such as the type and amount of other ions in the solution, may also effect the product distribution, for example, the type of catholyte and its concentration. In this work, during the electrochemical reduction of CO_2, a large peak at -0.98 V was observed, and it was confirmed that this peak was due to the electrochemical reduction of CO_2.

REFERENCES

[1] M. Gattrell, N. Gupta, A. Co, Journal of Electroanalytical Chemistry 594 (2006) 1–19

[2] S. Kaneco, H. Hiei , Y. Xing , H. Katsumata, H. Ohnishi, T. Suzuki, K. Ohta, Electrochimica Acta 48 (2002) 51-55

[3] M. Jitaru, Journal of Applied Electrochemistry 27 (1997) 875-889

[4] M. Watanabe, M. Shibata, A. Kato, J. Electrochem. Soc., 138 (1991) 528-531

[5] Y. Hori, K. Kikuchi, S. Zusuki, Chemistry Letters (1985) 1695-1698

[6] Y. Hori, K. Kikuchi, S. Suzuki, Chem. Lett. (1985) 1695

[7] Y. Hori, A. Murata, Electrochim. Acta 35 (11/12) (1990) 1777

[8] S. Taguchi, A. Aramata, Electrochim. Acta 39 (17) (1994) 2533

[9] H. Yoshitake, T. Kikkawa, K. Ota., Journal of Electroanalytical Chemistry, 390 (1995) 91-97

[10] D. Kolbe, W. Vielstich, Electrochimica Acta 41(1996) 2457-2460

[11] J.J. Kim, D.P. Summers, K.W. Frese, J. Electroanal. Chem. 245 (1988) 223

[12] Y. Hori, O. Koga, Y. Watanabe, T. Matsuo, Electrochimica Acta 44 (1998) 1389-1395

[13] Y. Hori, A. Murata, R. Takahashi, S. Susuki, J. Chem. Soc.,Chem. Commun. (1988)17

[14] Y. Hori, A. Murata, R. Takahashi, J. Chem. Soc., Faraday Trans. 1 85 (1989) 2309

[15] K. Ogura, H. Yano, and F. Shirai, J. Electrochem Soc., 150 (2003) D163

[16] H. Yano, T. Tanaka, M. Nakayama, K. Ogura, J. Electroanalytic chem., 565 (2004) 287

[17] S. Kaneco, N. Hiei, Y. Xing, H. Katsumata, H. Ohnishi, T. Suzuki, K. Ohta , Electrochimica Acta 48 (2002) 51-55

[18] H. De Jesús-Cardona, C. del Moral, C. R. Cabrera, H. De Jesus-Cardona, Journal of Electroanalytical Chemistry 513 (2001) 45–51

[19] G. Horányi, Journal of Electroanalytical Chemistry 453 (1998) 113–119

[20] G. Z. Kyriacou and A. K. Anagnostopoulos, J. Appl. Electrochem., **23**, 483 (1993) 267

[21] R. L. Cook, R. C. MacDuff, and A. F. Sammells, J. Electrochem. Soc., **134**(1987) 2375

[22] R. L. Cook, R. C. MacDuff, and A. F. Sammells, J. Electrochem. Soc., **135**(1988)1320

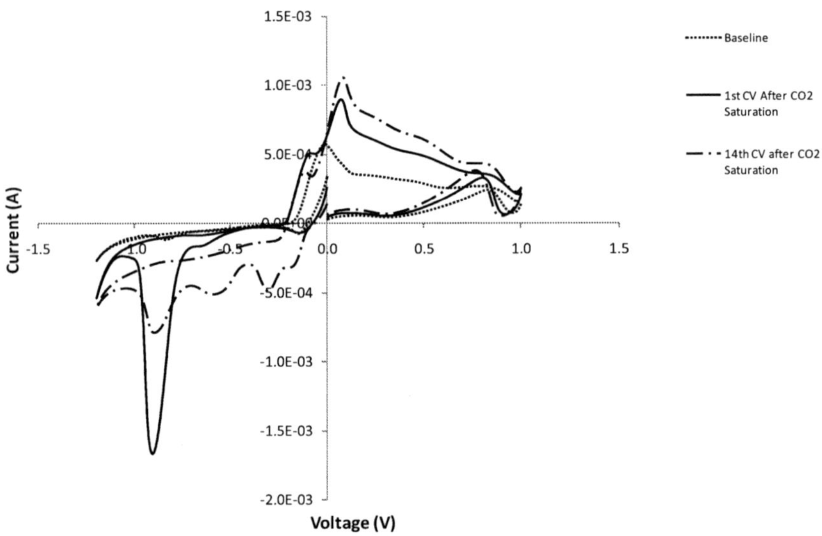

Figure 1. Cyclic Voltammetry tests on $KHCO_3$ solution during the electrochemical reduction of CO_2. Cathode: working (Cu wire), 20 mL of 2.0M $KHCO_3$, reference electrode (Ag/AgCl). Anode: 20 mL of 0.1M (100mM) H_2SO_4 and Counter electrode (Pt wire). This experiment includes 1hr CO_2 saturation after the CV for the base line was recorded.

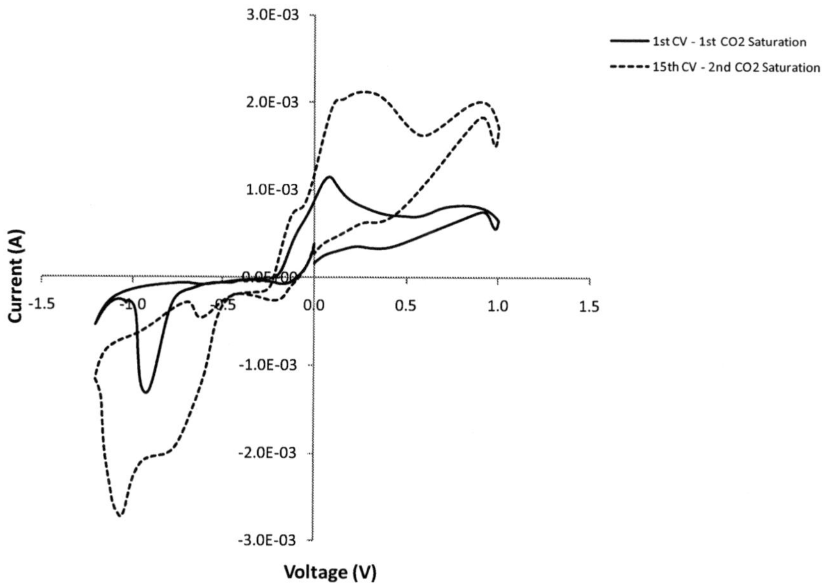

Figure 2. Cyclic Voltammetry tests in 2.0 M $KHCO_3$ solution during the electrochemical reduction of CO_2. The first CV was recorded after the solution was saturated with CO_2 for 1hr. Fourteen CVs were taken and then the solution was again saturated for 1 hr. Following, the 15th CV was recorded.

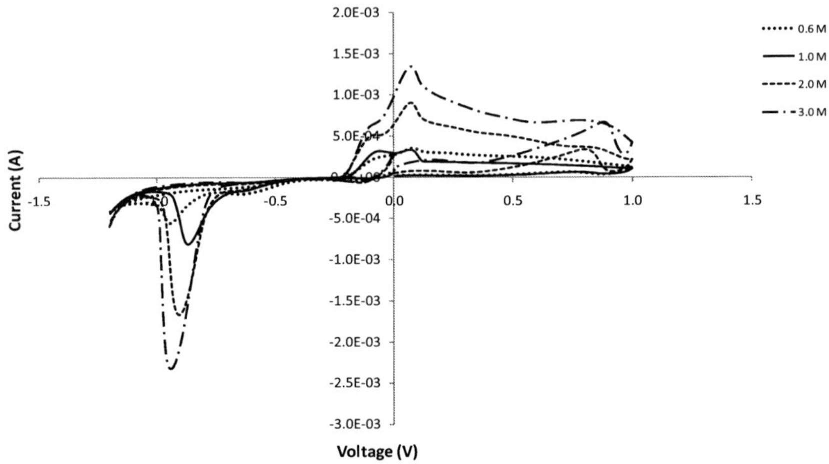

Figure 3. Cyclic Voltammetry tests with 0.6, 1.0, 2.0 and 3.0M $KHCO_3$ solutions during the electrochemical reduction of CO_2. Cathode: working (Cu wire), reference electrode (Ag/AgCl). Anode: 20 mL of 0.1M (100mM) H_2SO_4 and Counter electrode (Pt wire). This experiment includes 1hr CO_2 saturation after the CV for the base line was recorded.

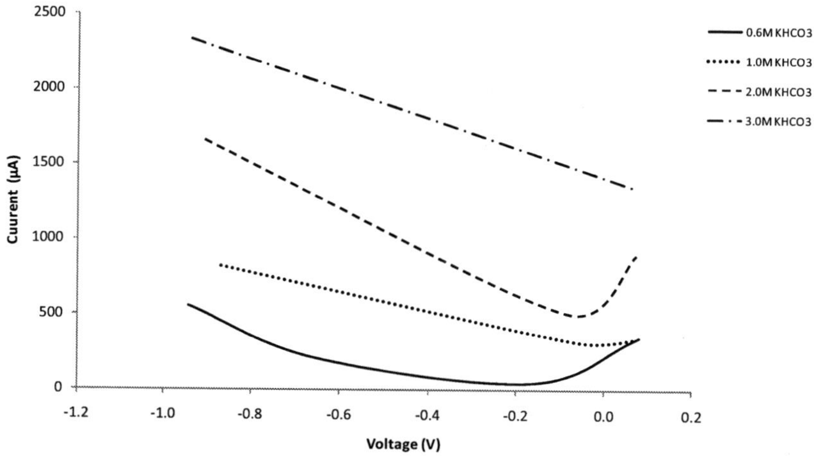

Figure 4. Current vs. Voltage recorded during the cyclic voltammetry tests with 0.6, 1.0, 2.0 and 3.0M $KHCO_3$ solutions during the electrochemical reduction of CO_2. Cathode: working (Cu wire), reference electrode (Ag/AgCl). Anode: 20 mL of 0.1M (100mM) H_2SO_4 and Counter electrode (Pt wire).

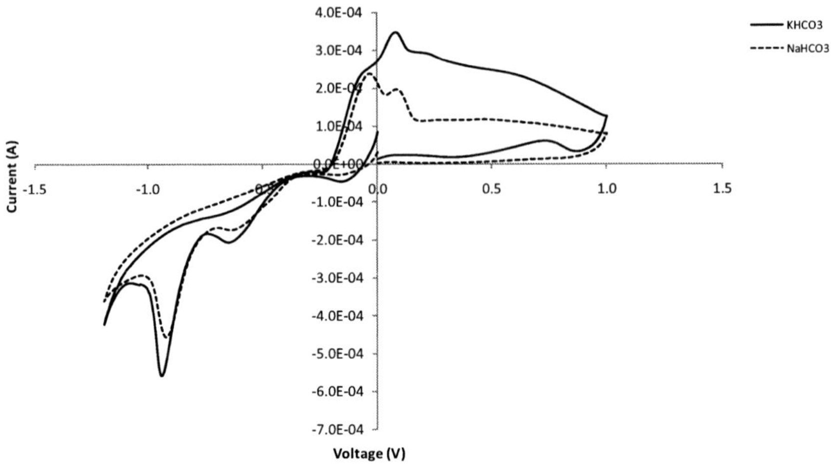

Figure 5. Cyclic Voltammetry tests with 0.6M $KHCO_3$ and 0.6M $NaHCO_3$ solutions during the electrochemical reduction of CO_2. Cathode: working (Cu wire), reference electrode (Ag/AgCl). Anode: 20 mL of 0.1M (100mM) H_2SO_4 and Counter electrode (Pt wire). This experiment includes 1hr CO_2 saturation after the CV for the base line was recorded.

Energy Technology 2010
Edited by: Neale R. Neelameggham, Ramana G. Reddy, Cynthia K. Belt, Ann M. Hagni and Subodh Das
TMS (The Minerals, Metals & Materials Society), 2010

THE THERMAL GAS PROCESSING IN PRE-HEATING ZONE OF "WATER-JACKET" FURNACES IN "TREPÇA".

A. HAXHIAJ[1], and E. HAXHIAJ[2]

[1]Faculty of Mining and Metallurgy, Mitrovica, University of Pristina; Republic of Kosovo, Peja, Raushiq, Postal code 30000
[2]American University in Kosovo, Republic of Kosovo, Peja, Raushiq, Postal code 30000

Keywords: load, coke, heat, humidity, dusts, gasses, environment.

Abstract

The paper has as a studying object the issue of effective management of gasses' heating and dusts in reductive melting process in lead (Pb) metallurgy. It also analyses the issue related to the usage and recycling of dusts and gasses in all pyro-metallurgical processes of technical lead production. The paper contains the analysis of technological process of separation of humidity, gasses, and dusts of valuable remainders in "Trepça" plant. Paper has the results of theoretical-experimental studying, analytical and graphical of gasses thermal balance depending on height and composition of load, and the coke quantity on load for reductive melt in "Water-Jacket" furnaces. With thermal balance, having into consideration all parameters of heat in entry and exit depending on aforementioned factors we aimed to optimize the parameters of the process, keeping the attention to the minimization of energy consumption, and the overall influence on environment.

1. Introduction

The upper part of the furnace which is the study point of this paper lies from the reactions zone of Water-Jacket furnace and it also includes the furnace neck. In this zone has been elaborated the zonal material balance referenced third in the row. The separation of humidity, dusts, gases and lead with gases depends on many parameters like temperature, load composition, and height of load inserted in the furnace, and also the height of preheating zone in Water-jacket furnace. Lead oxidizes in reaction zone of the furnace and exhales with gases like dusts and small particles after which falls in worthy storage of Trepça plant and has negative effects in environment, and economic stability of reductive melting process of Water-jacket furnace. Thermal balance of gases in this zone depends on many factors; height and load composition, coke quantity in a load. The calculation has been done as the standard methodology of thermal calculation is done; moreover, by using a computer for mathematical models which enables very accurate results. We must emphasize that acquired and used heat in reaction zone and melted materials is calculated in others studies, so their influence in preheating zone is assumed.

2.0 The Calculation of Thermal Balance in Preheating Zone of Water-Jacket Furnace

The thermal balance in preheating zone in Water-jacket furnace consists of acquired heat which can be variable and depends on the chemical composition of load, agglomerate and reductive-oxidative features of agglomerate. Simultaneously, thermal balance in preheating zone consists of used heat which also can be variable and depends on load height inserted for reductive melt, the quantity of coke inserted in furnace and the quantity of formed gases. These gases are exhaled in the atmosphere from the reductive melting process in Water-jacket furnace.

2.1 Acquired Heat

a) The heat that enters with agglomerate which is taken resources is calculated with formula:

$$Q_{agl} = G_{agl} \cdot C_{agl} \cdot t_{agl} (kJ/h)$$

Abbreviations of which are:

G_{agl} - agglomerate weight $(22250 kg/h)$,

C_{agl} - specific heat of agglomerate $(0.120 kJ/kg \cdot {}^{o}C)$,

t_{agl} - agglomerate temperature $(25^{o}C)$,

$Q_{agl} = 66750 kJ/h$

b) The heat that enters with agglomerate which immediately is inserted in furnace from the agglomeration machine type of Dwight Lloyd or Lurgi is calculated with the formula:

$$Q'_{agl} = G_{agl} \cdot C_{agl} \cdot t_{agl} (kJ/h)$$

Abbreviations of which are:

G_{agl} - agglomerate weight $(22250 kg/h)$,

C_{agl} - specific heat of agglomerate $(0.120 kJ/kg \cdot {}^{o}C)$,

t_{agl} - agglomerate temperature $(50^{o}C)$

$$Q'_{agl} = 133500 kJ/h$$

Table I shows acquired heat with agglomerate which depends on agglomerate temperature.

Table I. Acquired Heat with Agglomerate which Depends on Agglomerate Temperature

Agglomerate temperature (${}^{o}C$)	Agglomerate quantity (kg/h)	Specific heat of agglomerate ($kJ/kg^{o}C$)	Quantity of heat (kJ/h)
25	22250	0.120	66750
50	22250	0.120	133500

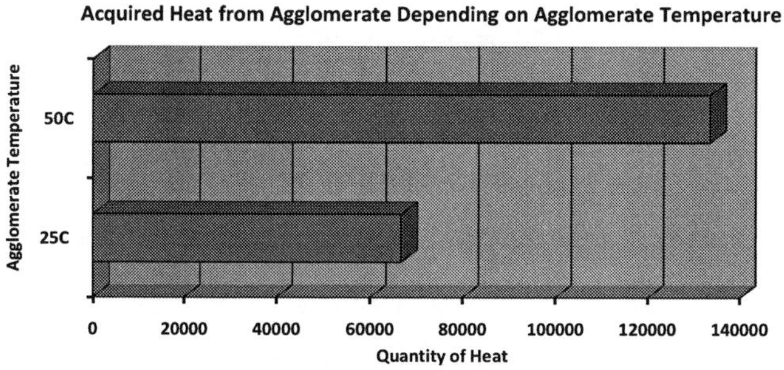

Fig. 1 Acquired heat from agglomerate depending on agglomerate temperature.

2.2 The Used Heat

For the sake of having exact management of thermal gases in preheating zone the calculation of used heat is done by alternatives based on enthalpy values.

a) The usage of heat to boil the humidity from coke is calculated with formula:

$$Q^{H_2O} = i^{H_2O} \cdot V(kJ/h)$$

$$i^{H_2O} = C_p^{H_2O} \cdot t(kJ/m^3)$$

i^{H_2O} - the enthalpy of steam,

$C_p^{H_2O}$ - specific heat of steam ($1.56 kJ/m^{3\,o}C$),

t - the separation temperature of crystal water (401°C,

$$i^{H_2O} = 626.2 kJ/m^3$$

$$Q^{H_2O} = i^{H_2O} \cdot V(kJ/h)$$

V - the quantity of humidity in a kilogram coke is calculated with formula:

$$V = R_t' - R_1 (m^3/kg)$$

R_t - the theoretical quantity of wet gases calculated with Rumumel formula:

$$R_t' = 1.5 + 0.5 \frac{Q_f}{1000} (m^3/kg)$$

Q_f - thermal power of coke.

$$R_t' = 31.2 m^3/kg$$

R_1 - theoretical quantity of dry gases taken from diagram J_t depending on Q_f and λ.

$$R_1 = 12 m^3/kg$$

$$V = 31.2 - 12 = 19.2 m^3/kg_{koks}$$

Table II shows the quantity of humidity in coke and the value of used heat for boiling the humidity calculated with enthalpy.

Table II. The Quantity of Humidity in Coke and the Value of Used Heat for Boiling the Humidity Calculated with Enthalpy

Coke quantity (kg)	Quantity of humidity (m^3/kg)	Enthalpy (kJ/m^3)
1	19.2	626.2
2	36.4	1252.4
3	57.6	1878.6

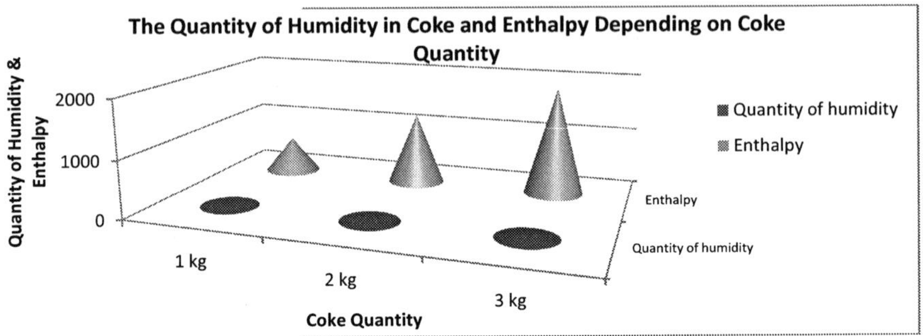

Fig. 2 Thecoke quantity, humidity, and the value of heat usage calculated with enthalpy.

b) The loss of heat with gases is calculated with formula:

$$Q^{gas} = i^{gas} \cdot V (kJ/h)$$

$$i^{gas} = C_p^{gas} \cdot t^{gas} (kJ/m^3)$$

i^{gas} - gases' enthalpy,

C_p^{gas} - specific heat of gases $1.36 kJ/m^3 {}^oC$

t^{gas} - gases' temperature on the exit of furnace neck which depends on many factors and has different values, in this case 300^oC.

$$i^{gas} = 408 kJ/m^3$$

Table III shows the enthalpy for gases depending on temperature of gases in the way out of furnace neck.

Table III. The Enthalpy for Gases Depending on Temperature of Gases in the Way out of the Furnace Neck

Specific heat of gases $(kJ/m^{3o}C)$	Temperature of gases (^oC)	Enthalpy of gases (kJ/m^3)
1.36	300	408
1.36	400	544
1.36	450	612
1.36	500	680

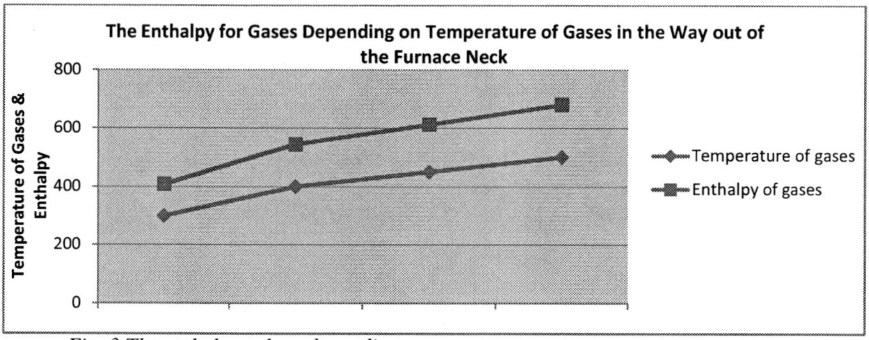

Fig. 3 The enthalpy values depending on gases temperature.

c) Volume of dry gases during coke burning is calculated proportionally as below:

$$1kg_{coke} \quad - \quad 12\,m^3/kg_{coke}$$

$$2420kg_{coke} \quad - \quad V$$

$$V = \frac{2420kg_{coke} \cdot 12\,m^3/kg_{coke}}{1kg_{coke}}$$

$$V = 29040\,m^3/kg_{coke}$$

d) The quantity of water in coke is the humidity percentage in coke, if there is 10% humidity in coke then water quantity in coke will be:

$$m_{H_2O} = G_{coke} \cdot 10\%(kg/h)$$

$$m_{H_2O} = 2420\,kg/h \cdot 0{,}1$$

$$m_{H_2O} = 242\,kg/h$$

Steam volume in gases is calculated proportionally as below:

$$1kg_{coke} \quad - \quad 19.2\,\mathrm{m}^3/\mathrm{kg}_{coke}$$

$$2420kg_{coke} \quad - \quad V$$

$$V = \frac{2420\mathrm{kg}_{coke} \cdot 19.2\,\mathrm{m}^3/\mathrm{kg}_{coke}}{1\mathrm{kg}_{coke}}$$

$$V = 46464\,m^3/kg_{coke}$$

e) Heat loss for boiling humidity from coke is calculated with formula:

$$Q^{H_2O} = i^{H_2O} \cdot V(kJ/h)$$

$$Q^{H_2O} = 626.2 \cdot 46464\,kJ/h$$

$$Q^{H_2O} = 29095756{,}8\,kJ/h$$

f) Heat loss with gases is calculated with formula:

$$Q^{gas} = i^{gas} \cdot V(kJ/h)$$
$$Q^{gas} = 11848320\, kJ/h$$

Table IV shows heat loss for boiling humidity from coke and heat loss with gases.

Table IV. The Loss of Heat for Boiling Humidity from Coke and Heat Loss with Gases.

Component	Enthalpy (kJ/m^3)	Volume of dry gases (m^3/kg)	Volume of gases with steam (m^3/kg)	Heat loss (kJ/h)
Boiling humidity from coke	626.2		46464	29095756.8
With gases which exit furnace	408	29040		11848320

3.0 The Quantity and Temperature of Gases Depending From Load Height Inserted in the Furnace

The preheating zone serves for several processes: to fill in the load, to dry the load, and to remove gases from furnace with variable quantity which depends on humidity quantity in load, coke and load height which is inserted to melt in Water-jacket furnace.
Table V show technologic indexes of melt with high and low pillar of load.

Table V. The Technologic Indexes of Melt with High and Low Pillar of Load.

Name	Unit	Melt with high pillar (5-6m)	Melt with low pillar (2.5-3m)
Capacity	t/m^2	45-60	60-75
Coke consumption	%	9-11	7.5-10
Pb quantity in sinter	%	Up to 1.5	3-3.5
Temperature of gases on furnace neck	^oC	100-150	600
Air consumption	m^3/t	900	1400
Quantity of dusts	%	0.5-1	3-5

a) Heat loss with gasses with low pillar of load for melt in Water-jacket furnace is calculated with:
   $$Q = R \cdot C_p \cdot t \cdot m(kJ/h)$$

R - Quantity of gasses which exit the furnace $(15.4\, m^3/kg)$
C_p - Specific heat of gases $(1.36\, kJ/m^3)$
t - Temperature of gases that exit the furnace (600^oC)
m - Consumed quantity of coke $(2420\, kg/h)$

$$Q = 30410688\, kJ/h$$

b) Heat loss with gases in high pillar of load for melt in Water-jacket furnace is calculated with formula:

$$Q^U = R \cdot C_p \cdot t \cdot m (kJ/h)$$

R - Quantity of gases that exit the furnace $(15.4 m^3/kg)$,

C_p - Specific heat of gases $(1.36 kJ/m^3)$,

t - Temperature of gases that exit the furnace (100^oC),

m - Consumed quantity of coke $(2420 kg/h)$,

$$Q^U = 5068448 kJ/h$$

Table VI shows heat loss with gases for high and low pillar of load.

Table VI. The Heat Loss with Gases for High and Low Pillar of Load.

Pillars of load	Quantity of coke consumed (kg/h)	Gases temperature ^oC	Quantity of lost heat (kJ/h)
Low pillar	2420	600	30410688
High pillar	2420	100	5068448

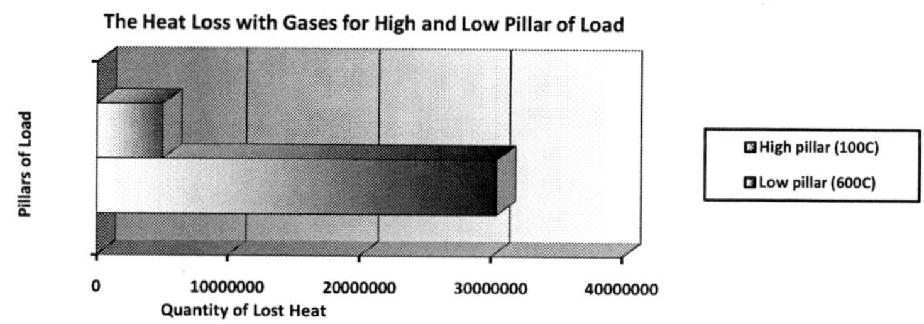

Fig. 4. The heat loss with high and low pillar of load for melt in Water-jacket furnace in "Trepça".

4. Conclusion

Having considered the calculation we can conclude that heat lost with gases has high values, in particular for melting process with low pillar of load in the furnace. Temperature of gases mainly depends on height of load and load composition for melt, quantity of consumed coke per ton technical lead produced in Water-jacket furnaces in Trepça.

Based on analytical and experimental calculation of thermal balance of gases in preheating zone of Water-jacket furnace we can argue that acquired energy (that enters the process) is minimal comparing to used energy.

Thus, this difference results in emission formation, dusts layers, and gases that convey heat and dusts on environment which pollute the environment and destabilize reductive melting process of agglomerate in Water-jacket furnace.

In conclusion, to reduce the emission formation, gases, dust layers, and loss of heat quantity with gases we should influence in:

- The enlargement of load height in furnace,
- Exact consumption of coke per ton technical lead produced,
- Load consumption for melt with minimal humidity, and
- Exact consumption of air for coke combustion.

References

[1]. Fatmir Agolli, *Metalurgjia e Metaleve me Ngjyrë* (Prishtinë, Universiteti i Prishtinës, 1985), (262-270).

[2]. Miodrag Spasić, and Dushan Vučurović, *Teorija Metalurških Procesa – I deo, sveska 2,* (Beograd, Tehnološko-Metalurški Fakultet, Univerziteta u Beogradu, 1982), (4-10).

[3]. A. Haxhiaj, "Materialna i toplotna bilansa PORT-PIRI peći" (Master thesis, Sveučilište u Zagrebu, 1989), (42-48).

[4]. A. Haxhiaj, "Intensifikimi dhe Optimalizimi i Procesit të Shkrirjes Reduktuese të Aglomeratit të Plumbit në Furrat Shahte" (Ph.D. thesis, Universiteti i Prishtinës, 2003), 37-41.

[5]. Hans-Joachim Bittrich, *Leitfaden der Chemischen Thermodynamic* (Berlin, 1976), 93-114.

[6]. A. Haxhiaj, D. Elezi, and Z. B. Gashi, Bilanci termik në zonën e reaksioneve të furrave "PORT-PIRI" (Prishtinë, ACTA CHIMICA KOSOVICA, 2005), 85-96.

[7]. M. Jovanović, "Nova tehnologija i postrojenja za proizvodnju-olova", (Paper presented at the Zbornik radova II Jugoslovenskog simpozijuma o metalurgij, TMF, Beograd, Serbia, 1978), 8, 31, 51.

[8]. B. Vasiljević, "Tendencije razvoja metalurgije olova u svetu i kod nas" (Paper presented at the Zbornik radova i Jugoslovenskog simpozijuma o metalurgiji, TMF, Beograd, Serbia, 1974), (187-210).

[9]. A. Haxhiaj, M. Rizaj, N. Murati, and Sh. Ibrahimi, Bilanci i nxehtësisë në zonën e materialit të shkrirë në furrat Vaterxhakete, (Prishtinë, Akademia e shkencave dhe e arteve e Kosovës, Seksioni i shkencave të natyrës, Kërkime 12, 2004), 145-153.

Energy Technology 2010
Edited by: Neale R. Neelameggham, Ramana G. Reddy, Cynthia K. Belt, Ann M. Hagni and Subodh Das
TMS (The Minerals, Metals & Materials Society), 2010

OXIDATION KINETICS OF Fe-Cr AND Fe-V LIQUID ALLOYS UNDER CONTROLLED OXYGEN PRESSURES

Haijuan Wang[1)], Nurni N. Viswanathan[2)] and Seshadri Seetharaman[1)]

1) Division of Materials Process Science, Royal Institute of Technology, Stockholm, Sweden

2) Department of Metallurgical Engineering and Materials Science, Indian Institute of Technology Bombay, Mumbai, India

Keywords: Fe-Cr melt, Fe-V melt, Oxidation, CO_2

Abstract

In order to make the stainless steel making process efficient and environment friendly, it is essential to minimize the loss of chromium to the slag phase. With a view to investigate the advantages of using CO_2 to attain moderate oxygen partial pressures in the oxidant gas during the decarburizations of stainless steel, the present experiments were carried out to understand the oxidation kinetics of Fe-Cr alloys and Fe-V alloy with gas mixtures containing CO_2.

In the present work, the oxidation behavior of Fe-Cr and Fe-V was examined by thermogravimetric analysis (TGA). Various mixtures of O_2 and CO_2 were used as the oxidant gas and the oxidation rate was followed by the weight changes as a function of time.

The experiments demonstrated that the oxidation rate is independent of temperature at present experimental situation, but has strongly related to the Cr-content of the alloy as well as the oxygen partial pressure in the oxidant gas mixture. The wetting of alumina by the iron drop and the change of the drop shape during the course of the oxidation were investigated by X-ray radiography.

1. Introduction

During the production of high alloy steel process, it is essential to minimize the loss of chromium to the slag phase in view of the economic and environmental impacts. In this case, the study of mechanism on the oxidation process of Fe-Cr alloy has attracted a great deal of interest among materials scientists. Rhys-Jones *et al.* [1] have studied the isothermal-oxidation kinetics of solid Fe-10%Cr alloy at 1273 K and has shown that the oxidation rate follows a parabolic rate law. It is also pointed out by Rhys- Jones *et al.* [1] that during the oxidation of Fe-20%Cr alloy for 70hours, the products were identified as Fe_2O_3 and Fe_3O_4. Wood *et al.* [2] reported that there is a spallation oxidation of Fe-Cr alloys, and the oxidation rate shows a quick increase after a protective induction period. As expected, only Cr_2O_3 in the case of the alloy with 14-25% Cr. Mortimer and Sharp [3] reported that the oxidation of alloys with high Cr content results in the formation of the oxide solid solution $(Cr, Fe)_2O_3$ in the product layer. Holappa and Forsbacka [4] have proved, the higher oxygen partial pressure, the more $Cr^{3+}(CrO_{1.5})$ is present in the slag and the melting point is increased. The equilibrium oxide phases with liquid Fe-Cr alloys were studied by Hilty *et al.* [5] at 1823K, 1873K and 1923K. These authors point out that, the product layer consists of, besides chromite($FeO{\cdot}Cr_2O_3$) and Cr_2O_3, two unknown oxide phases has existed, one is Cr_3O_4 with tetragonal crystal structure, the other is a distorted spinel with the composition between $FeO{\cdot}Cr_2O_3$ and Cr_3O_4. Richards and White [6], Woodhouse and White [7] have also studied the equilibria in the Fe-Cr-O system from 1693K to 1923K under the air atmosphere. Muan *et al.* [8] illustrated the phases in Fe-Cr-O system from 1673K to 2323K in air.

It is seen that the earlier investigations on the oxidation of Fe-Cr alloys were mostly focused on the solid state oxidation in oxygen or air. To the knowledge of the present authors, no study on the oxidation of liquid Fe-Cr alloys under different oxygen partial pressures has so far been reported. Knowledge of the kinetics of oxidation of Fe-Cr alloys in CO_2 is extremely relevant to Cr-retention in stainless steel when decarburized with CO_2 investigated by the present authors earlier [9]. Thus, the present work was carried out in order to understand the mechanism of oxidation Fe-Cr alloys with CO_2 as well as mixtures of CO_2 and O_2.

2. Experiment Set up

2.1 Materials

Iron powder with a purity of +99.9% (<10micro, Sigma Aldrich, Germany) and Chromium powder with a purity of 99%(Sigma Aldrich, Germany) were used in the present experiments. These two metal powders were mixed with a horizontal roll mixer for 24 hours before experiment. The materials were then pressed into pellets with diameter of 6mm and hold in an alumina crucible. The pellets were weight accurately before every investigation.

2.2 Experimental Apparatus and Procedure

The oxidation process of Fe-Cr melt oxidation was investigated using Thermogravimetric Analysis (TGA). A SETARAM TAG-92(France) thermogravimetric instrument was employed for this purpose, a schematic picture of which is shown in Figure 1. The experimental apparatus consists of an electronic microbalance, a graphite heating furnace, an Al_2O_3 reaction tube and a computer control penal. The instrument is also equipped with an accessorial gas cleaning system, which has been presented [10,11] in the earlier publications of the present research group.

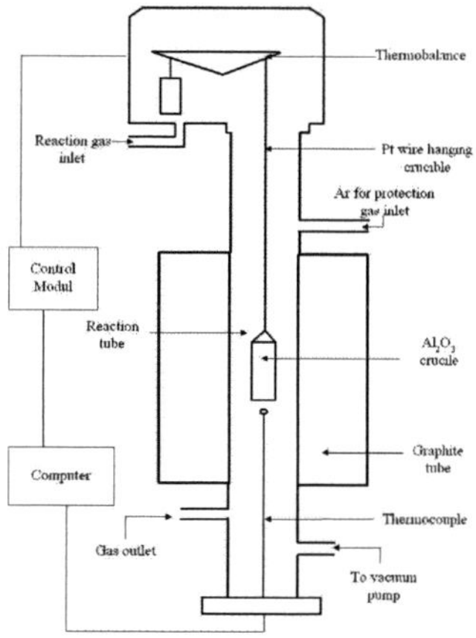

Figure 1. The schematic of experimental set up.

An alumina crucible with O.D. 10mm, I.D. 8mm and height 15mm, in the dense of 99.5% Al_2O_3 was used to hold the samples. 400mg the alloy was used in each experiment. The crucible was placed in a Pt basket and hung by a Pt suspension wire from the thermobalance. The basket was introduced into the alumina reaction tube (with I.D. of 16mm) which was positioned inside the graphite furnace as shown in Figure 1. The length of Pt wire for experiments was adjusted carefully so that the sample was located in the even temperature zone of the furnace. A thermocouple was positioned close to the bottom of the alumina crucible. The reaction tube was first evacuated for 180s down to a vacuum of 10 Pa and then flushed with purified argon gas through the reaction gas inlet shown in Figure 1. The sample was then heated to the targeted temperature with the maximum rate of 15K/min with a constant flow rate of protective argon gas, which was purified of oxygen impurity so that the oxygen partial pressure in the gas was lower than 10^{-15} Pa. No weight gain could be noticed during this heating period confirming thereby that the purity of the argon gas was quite satisfactory. After the attainment of the targeted temperature, the temperature was maintained for about 1.5 h so that the sample attained thermal equilibrium. The oxidant gas was then introduced and the oxidation reaction was started. The weight changes of samples were recorded every 4s by the computer during the experiments. At the completion of the oxidation, the oxidant gas was replaced by Ar gas and the sample was cooled at rate 20K/min. The reaction products were ground into powders and sent to detect with X-Ray Diffraction (XRD), and chemical analysis. A list of the experiments conducted is presented in Table 1.

Table 1. Details of the Present Experimental Series

	Materials	Temperature(K)	Gas	Gas flowrate (Nml/min)
1	Fe+11wt%Cr	1873	100% Ar(purified)	150
2	Fe+11wt%Cr+3.5wt%C	1873	100% Ar(purified)	150
3	Fe+11wt%Cr	1873	100% CO_2	50
4	Fe+11wt%Cr	1873	100% CO_2	150
6	Fe+11wt%Cr	1873	100% CO_2	250
8	Fe+11wt%Cr	1823	100% CO_2	150
9	Fe+11wt%Cr	1923	100% CO_2	150
10	Fe+16wt%Cr	1873	100% CO_2	150
11	Fe+21wt%Cr	1873	100% CO_2	150
12	Fe+11wt%Cr	1873	36%CO_2+9%O_2+55%Ar	150
13	Fe+11wt%Cr	1873	80%CO_2+20%O_2	150
14	Fe+11wt%Cr	1873	60%CO_2+40%O_2	150
15	Fe+11wt%Cr+1.5wt%C	1873	100% CO_2	150
16	Fe+11wt%Cr+2.5wt%C	1873	100% CO_2	150
17	Fe+11wt%Cr+3.5wt%C	1873	100% CO_2	150

3. Results

3.1 The Products of Oxidation

XRD and chemical analysis of the products from the various oxidation experiments revealed that the products differed as the oxygen partial pressure of the oxidant gas was varied. With 20% or 40% O_2 in the oxidant gas, the products were distorted spinel (FeO dissolved in $FeCr_2O_4$) and Cr_2O_3. In the case of the gas mixture 36%CO_2+9%O_2+55%Ar, the spinel phase (FeO dissolved in $FeCr_2O_4$), Fe_3O_4 and $(Cr, Fe)_2O_3$ were found to be in the product layer. These are in accordance with the report by Mortimer and Sharp [3]. With pure CO_2, the product was found to be mainly FeO dissolved $FeCr_2O_4$ spinel. It is found that, higher oxygen partial pressure in the oxidant gas leads more Cr^{3+}(Cr_2O_3) in the slag, which is in accordance with the result reported by Holappa et al..[4]

3.2 The Determination of Gas Starvation Flow Rate

For the present experiments, a gas flow rate of 150Nml/min for the oxidant gas flow was used in the case of the tests. This flow rate was chosen after conducting oxidation experiments at 1873K with the same composition of materials of Fe+11%Cr in order to determine the starvation rate so that the gas phase mass transfer will not have any impact on the experimentally observed rate of the reaction.

3.3 Effect of Temperature on the Oxidation

The compositions of the alloys used in the present experiments are marked in the phase diagram presented in Figure 2.

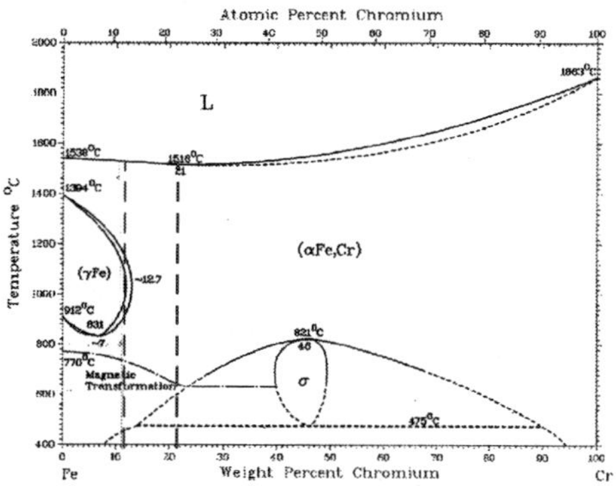

Figure 2. The compositions selected in the present experiments.

When the chromium contents are in the range of 11wt%-21wt%, according to the published phase diagram of Fe-Cr system [12], the Fe-Cr alloys studied in the present investigation have liquidus temperatures above 1811K. The oxidation rates were studied at three different temperatures, *viz.*1823, 1873 and 1923K. The thermogravimatric results in the case of the oxidation of Fe+11% Cr in pure CO_2 for the three temperatures is presented in Figure 3.

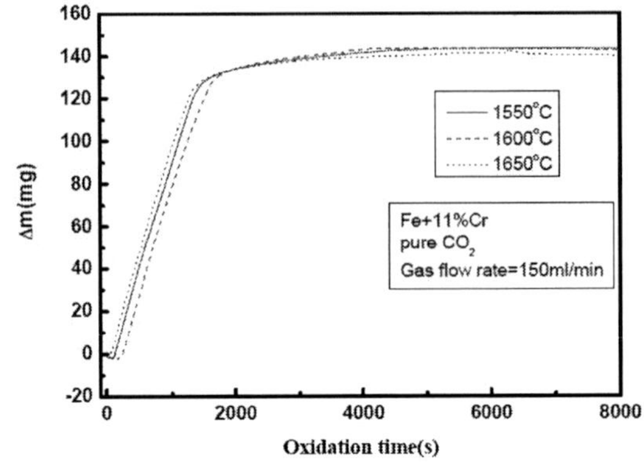

Figure 3. The weight change at different temperatures for Fe+11%Cr melt in contact with pure CO_2.

From Figure 3, it can be seen clearly that, for the same material, the weight gain in the case of oxidation with pure CO_2 is almost the same at all the three experimental temperatures. This indicates that temperature has no significant influence on the oxidation of Fe-Cr melt in the range of 1823K-1923K. XRD analysis of the reaction product indicated that the oxide product formed at 1823 and 1873 K were the same, *viz.* distorted spinel (FeO dissolved in $FeCr_2O_4$). At 1923 K, Fe_3O_4 was also found besides the spinel phase.

3.4 Effect of Cr Content on Oxidation Rate of Fe-Cr Melt

Three different Fe-Cr alloys with Cr contents 11, 16 and 21wt% were used to study the effect of chromium content on the oxidation rate of the melt. The results at 1873 K are presented in Figure 4.

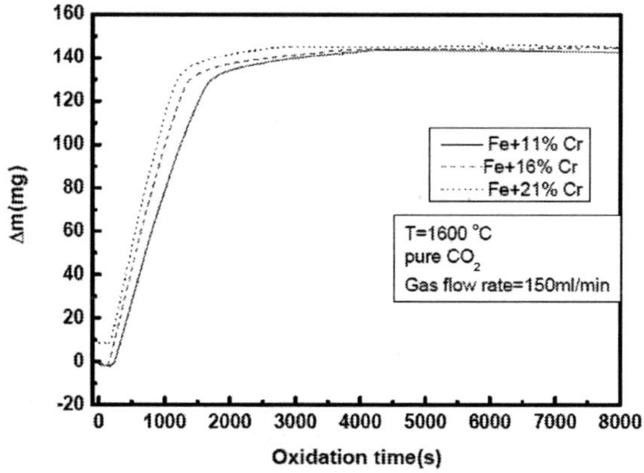

Figure 4. The effect of chromium content on the oxidation of Fe-Cr melts at 1873 K.

It can been seen in Figure 4 that, the oxidation rates are quite similar for all the three compositions even though there is a slight increase in the oxidation rate with increasing Cr content. The duration for the oxidation reaction reaching a steady state condition decreases slightly with increasing Cr content of the alloy: 1400 s for the alloy with 11% Cr, 1200 s for the alloy with 16% Cr and 1000 s for Cr content of 21%.

3.5 Effect of Oxygen Partial Pressure on the Oxidation Behavior of Fe-Cr Melt

The impact of the oxygen partial pressure in the oxidant gas on the oxidation process was investigated by carrying out the experiments in different O_2-CO_2 mixtures of varying oxygen partial pressures at 1873 K in the case of Fe-11%Cr alloy. The flow rate of 150ml/min was kept constant in al cases. The gas mixtures used were 100%CO_2, 36%CO_2+9%O_2+55%Ar, 80%CO_2+20%O_2 and 60%CO_2+40%O_2. The thermograms of these experiments are presented in Figure 5.

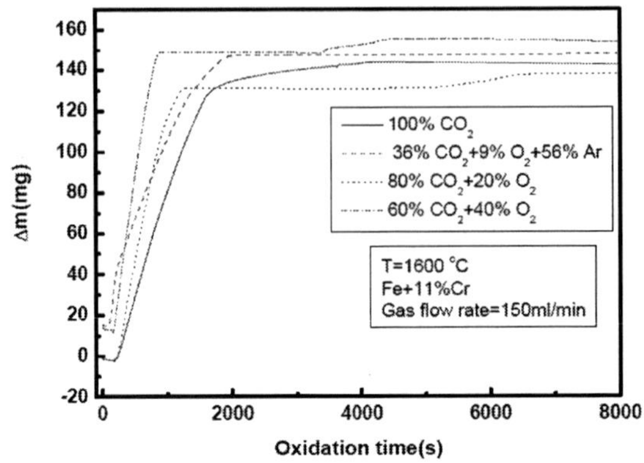

Figure 5. The effect of partial pressure of oxygen on the oxidation of Fe-Cr at 1873 K.

It can be read from Figure 5 that the oxidation rate increases with the increase of O_2 in the gas mixture, i.e. higher partial pressure of oxygen leads to quicker reaction, which is in accordance with the thermodynamic analysis. Besides, in the case of the oxidant 60% CO_2+40% O_2, the time for the attainment of steady state is much less compared to the other oxidant gases, around 15min; in the case of the gas mixture 60% CO_2+20% O_2, the duration to steady state is about 22min.; introduction of 9%O_2 plus 55% Ar for diluting the CO_2, increases the corresponding time to about 33min. In the case of pure CO_2, it takes around 60min to get final state. There is an interesting phenomena that, the weight change curve corresponding to the oxidant gas mixture 36 %CO_2+ 9%O_2+55% Ar shows an initial high oxidation rate, which subsequently becomes slower compared to the other gas mixtures except pure CO_2. The reason for this is uncertain. One plausible explanation can be that the formation of an initial passive layer through which the oxygen from the gas phase has to pass through is not covering the metal bath surface completely in this case leading to faster chemical reaction. With the continuation of the oxidation, the reaction slows down subsequently due to insufficient oxygen reaching the metal surface. Further experiments are needed to confirm this hypothesis.

3.6 The Effect of Carbon on the Oxidation of Liquid Fe-Cr System

It is important to find out the effect of carbon on the oxidation of Fe-Cr alloys as this will be highly relevant to the industrial process of decarburization of stainless steel. Three carbon contents were tested, *viz.* 1.5, 2.5 and 3.5wt%. The oxidation experiments were conducted at 1873 K. In this series, the oxidation was carried out in an atmosphere of pure CO_2, the gas flow rate being 150 ml/min. The result is illustrated in Figure 6.

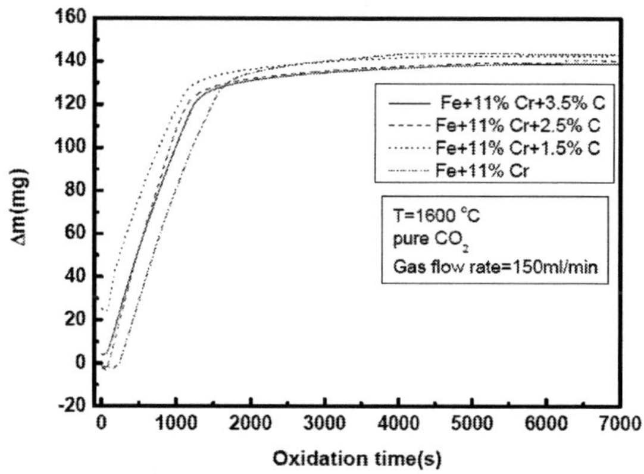

Figure 6. The effect of carbon on the oxidation of the Fe-Cr melt at 1873K.

The results of the experiments shown in Figure 6 show that the addition of carbon in Fe-Cr system increases the reaction rate, as represented by the shorter time interval before reaching the horizontal portion in the curves. However, with respect to the Fe-Cr-C melt, the alloy with 3.5%C has the lowest oxidation rate but the melt with 1.5% C get the higher oxidation rate, *viz.* the oxidation rate decreases with the increase of carbon content. It is to be remembered that the oxidant gas in this case oxidizes not only Cr but also C simultaneously as experienced by the present authors in a parallel series of experiments [9]. Thus, the weight gain recorded corresponds to the weight increase due to the oxidation of Cr and Fe minus the weight loss corresponding to carbon loss. One plausible explanation is the oxygen penetration of the metal bath and the formation of CO bubbles due to decarburization reaction:

$$\underline{C}_{Fe} + CO_2(g) = 2\,CO\,(g) \tag{1}$$

The formation of two CO molecules from one mole of CO_2 according to this reaction will lead to an increased stirring near the surface leading to more intensive oxidation of Cr from the bath. Another factor to be considered is the interaction parameter of Cr with respect to carbon in molten iron, ε_{Cr}^{C}, which is highly negative. Thus, it would be expected that the oxidation of Cr from the bath would be retarded at high carbon contents. Since this argument is contrary to the experimental observations, further work is carried out in this respect.

4. Discussions

The alloy compositions chosen in the present experiments are marked in Fe-Cr-O phase diagram at 1573K. The corresponding oxidation path can be traced along the dashed lines in Figure 7. With different Cr contents, i.e. 11%Cr, 16%Cr and 21%Cr, the product is the same, viz. distorted spinel (FeO dissolved in $FeCr_2O_4$), corresponding the points on the FeO-$FeCr_2O_4$ tie-line in Figure 7. With the O_2 ratio in gas mixture enhanced to 20%, Cr_2O_3 got formed, the product would lie in the three-phase region Fe_3O_4-$FeCr_2O_4$- Cr_2O_3, as shown in the figure.

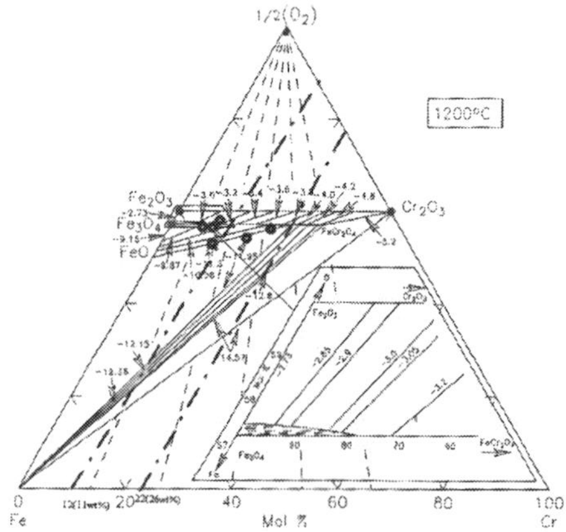

Figure 7. The products in the present experiments corresponding to the Fe-Cr-O phase diagram at 1573K [13].

XRD and chemical analysis indicated the partial oxidation of Fe from the metal bath, which again is expected as Cr is getting depleted in the melt, the Fe oxidation and the formation of the compound oxides that would lower the Gibbs energy of Fe in the product would be expected from a thermodynamic view point.

The following oxidation reactions at 1873K can be considered to take place during the present experimental series:

$$2\,\underline{Cr} + \frac{3}{2}O_2(g) \;=\; Cr_2O_3\ (s) \tag{2}$$

$$Fe + \frac{1}{2}O_2(g) + Cr_2O_3\ (s) \;=\; FeO\cdot Cr_2O_3\ (s) \tag{3}$$

$$Fe + \frac{1}{2}O_2(g) \;=\; FeO\ (l) \tag{4}$$

$$3FeO + \frac{1}{2}O_2(g) = Fe_3O_4\ (l) \tag{5}$$

In the initial stages of the process, the reaction (2) in the sequence above would be expected to occur leading to the formation of Cr_2O_3(s) on top of the melt. At this stage, the chemical reaction between the melt and the oxygen in the oxidant gas would essentially be the only reaction that would determine the rate.

In order to understand the phenomena at the interface, the following confirmative experiment was conducted. 400mg Fe-11%Cr mixed powder was melted under the purified Ar gas and quenched in the same atmosphere to get the figure of the melt before oxidation. It was observed that the sample before blowing oxidation gas is like a droplet, as shown in Figure 8. The radius of the droplet is about 4mm.

Figure 8. Fe+11%Cr melted under protection of Ar at 1873K.

On the surface of liquid Fe-Cr droplets, the following reactions between the liquid-gas are assumed to take place simultaneously with blowing pure CO_2.

$$CO_{2(ads)} \rightarrow CO_{(ads)} + O_{(ads)}$$ (6) [14,15]

$$2\underline{Cr} + 3\underline{O} + 3CO(gas) \rightarrow Cr_2O_3(s) + 3CO(gas) \tag{7}$$

$$Fe + \underline{O} + CO(gas) \rightarrow FeO(l) + CO(gas) \tag{8}$$

In order to follow the wetting characteristics as the oxidation reaction proceeds, the iron drop was placed in a high temperature X-ray radiography unit. A description of the apparatus is presented elsewhere[16]. The drop was heated in an atmosphere of purified argon with the oxygen partial pressure below 10^{-13} Pa. After the attainment of thermal equilibrium, the sample was exposed the oxidant gas at a flow rate of 150Nml/min.and the change in the shape of the drop was followed by X-ray photography taken every 10 min. with the help of the CCD camera fitted to the equipment in the dynamic mode as the oxidation proceeded. The experiment was conducted in the case of Fe+11%Cr alloy (800mg) at 1873K. These photographs are shown in Figure 9.

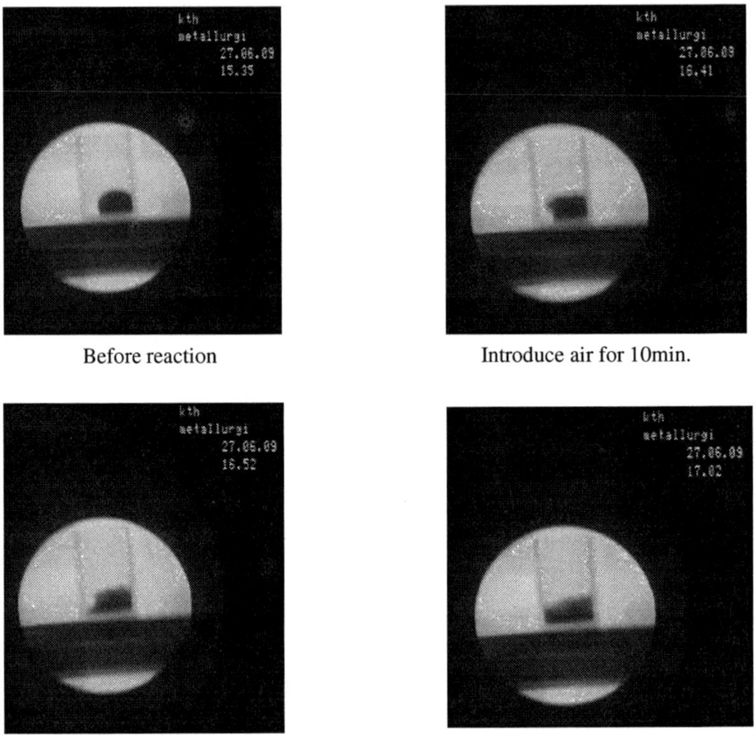

Before reaction

Introduce air for 10min.

Introduce air for 20min

Introduce air for 30min

Figure 9. The oxidation process of Fe+11%Cr at 1873K under air atmosphere.

As seen in Figure 9, the shape of the melt changed gradually with the oxidation process. Before reaction, Fe-Cr melt was as introduced had the shape of a drop; after oxidation for 10min, the top surface of the drop is enlarged; the material is molten and flows to the bottom after 20min reaction, and finally, the contents of the crucible form a liquid with a nearly horizontal surface. The various stages of oxidation as presented in Figure 10 may be describes as follows:

(1) The initial state after melting under Ar atmosphere. A droplet sits in the middle of the crucible bottom without contact with the crucible wall.
(2) The solid Cr_2O_3 layer getting formed on top of the droplet. At the initial stage of oxidation, the Cr is oxidized in priority and forming Cr_2O_3 particles. Those particles will float and sit on the surface of the droplet.
(3) The liquid FeO getting formed and combined with Cr_2O_3. FeO will be liquid at 1873 K and Cr_2O_3 then dissolves inside it. The liquid slag would spread to the gap between metal droplet and the crucible.
(4) Metal amount decreases with the oxidation, simultaneously, slag forms more. That would make the slag line and the metal line nearly to horizontal.

It can be assumed that, as the reaction proceeds, the liquid oxide layer is likely to cover the surface and the reaction rate may move towards diffusion control. The various steps involved in the reaction can be considered to consist of:

1) The diffusion of gas through the gaseous boundary layer to the gas-liquid slag interface.

2) The oxidant gas passes through the slag layer to the metal-slag interface.

3) Mass transfer of Cr through liquid bulk to the slag-liquid metal interface.

4) The chemical reaction at the droplet-slag interface.

5) Dissolution of the product oxide in the slag layer

6) The generated gas CO diffuses to gasous phase through slag layer.

In order to investigate which of these steps could be rate-controlling, further investigations regarding these steps are required.

The diffusion of the gas to the slag-metal interface: From the starvation rate studies, it is presumed that the gas phase mass transfer may not be important in determining the reaction rate. The transfer of oxygen from gas-slag interface to the metal-slag interface would be an ionic transport (O^{2-} ions) with electrochemical charge transfer. With the slag phase containing cations of variable valences, as for example Cr^{2+}/Cr^{3+} or Fe^{2+}/Fe^{3+}, the transport of electrons along with the O^{2-} ions would be favor the reaction.

Mass transfer of Cr from bulk to the metal-slag interface: The mass transfer would involve convective and diffusive mass transfer. The present results indicate that higher chromium content in the alloy leads to a faster oxidation. This indicates that the oxidation process is limited by the mass transfer of Cr from the liquid bulk to the interface. The diffusion rate is proportional to the difference in the chemical potential of Cr between the bulk and the interface. While the convective forces should be negligible in the present experimental set-up, there could be slight temperature differences in the melt as the exothermic oxidation occurs at the surface. Consequently, the surface would be at a higher temperature.

Reaction steps (5) and (6) can be considered as quite fast at the experimental temperature. The fine product particle formed in situ will dissolve in the slag layer immediately and the escape of CO gas in the case of the carbon-containing alloys can be considered as occurring quite fast.

As mentioned earlier, the chemical reaction at the slag metal interface would be rate-determining at the early stages of the oxidation process. As the product slag layer gets formed, the diffusion steps or mixed control could have a greater impact on the reaction rate. From the thermograms, it is seen that the weight change is mostly a straight line before the steady state conditions are reached. This suggests that even at later stages, the role of the chemical reaction is very important in determining the over-all reaction rate.

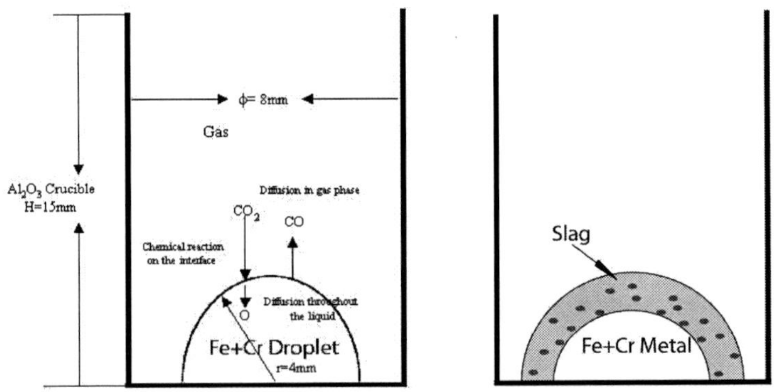

(1) Start oxidation (2) Solid oxide layer formed

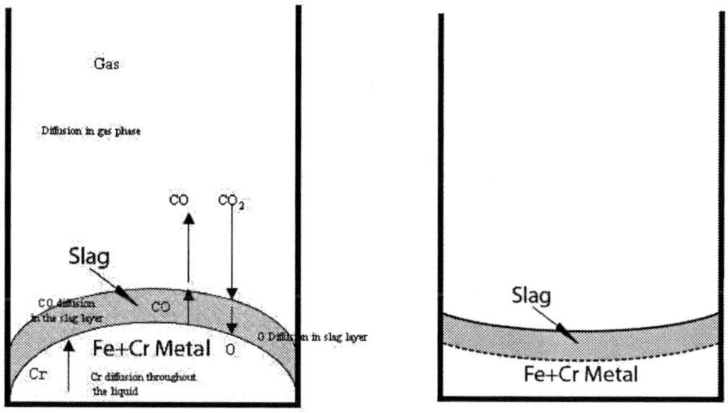

(3) Liquid slag layer shown up (4)Finish oxidation

Figure 10. The assumed oxidation kinetics steps of Fe-Cr melt under CO_2 atmosphere.

As mentioned before, the oxidation behavior of Fe-Cr melt strongly related to the partial pressure of oxygen. The evolution of processes for refining stainless steel, like AOD is, in fact based on reducing the oxygen partial pressure to the necessary and sufficient levels. The present authors [9] have shown that gas mixtures containing CO_2 would be very effective in the case of decarburization of Fe-Cr liquid alloys containing C resulting in a higher Cr retention than the conventional methods.

5 Conclusions

The oxidation behavior of Fe-Cr alloy in liquid state was investigated at 1823-1923K with varying oxygen partial pressure, and Cr content, from 11wt% to 21wt%. The oxidation rate is independent of reaction temperature under the present experimental situation, but is strongly related to the chromium content in the alloy. The higher chromium contents as well as higher oxygen partial pressure in the oxidant gas leads to a faster oxidation rate of the alloy.

The changes occurring due to the changing of the wetting of the liquid metal were investigated by X-ray radiography. The kinetic aspects of the oxidation reaction are examined and the importance of the chemical reaction is pointed out.

Acknowledgements

The authors wish to acknowledge Swedish Foundation for Strategic Environmental Research (MISTRA) through Jernkontoret, Sweden for partial financial support. China Scholarship Council kindly offers scholarship to Ms. Wang.

References

1. T. N. Rhys-Jones, H. J. Grabke, and H. Kudielka, *Corros. Sci.*, 27 (1987), 347.
2. G. C. Wood, T. Hodgekiess, and D. P. Whittle, *Corros. Sci.*, 6 (1966), 129.
3. M. Mortimer and W. B. A. Sharp, *Brit. Corros. J.*, 3 (1968), 61.
4. L. Holappa, L. Forsbacka and Zh. Han, "Measuring and Modeling of Viscosity and Surface Properties in High Temperature Systems," *ISIJ International*, 46 (3) (2006), 394-399.
5. D.C. Hilty, W. D. Forgeng and R. L. Folkman, *J. of Metals*, 7 (1955), 253-268.
6. R.G. Richards and J. White, *Trans. Brit. Ceram. Soc.*, 53 (1954), 422-459.
7. D. Woodhouse and J. White, *Trans. Brit. Ceram. Soc.*, 56 (1957), 569-580.
8. A. Muan and S. Sōmiya, *J. Am. Ceram. Soc.*, 43 (1960), 204-209.
9. H. Wang, N. N. Viswanathan, N. B. Ballal and S. Seetharaman, sent to HTMP for publication.
10. M. Persson, Sridhar Seetharaman and S. Seetharanman, *ISIJ International*, 47 (12) (2007), 1711-1717.
11. P. Dong, X. Wang, S. Seetharaman, *Steel Research International*, 80 (3) (2009), 202-208.
12. E. Schürmann, J. Brauckmann, *Arch. Eisenhüttenwes*, 48 (1977), 3.
13. R. Snethlage and D. D. Klemm, N. Jahrb, *Mineral., Abh.*, 125 (1975), 227-242.
14. D. R. Sain, G. R. Belton, *Metallurgical Transactions B (Process Metallurgy)*, 7B (2) (1976), 235-244.
15. A.W. Cramb and G. R. Belton, *Metallurgical Transactions B*, 12B (1981), 699-704.
16. G. Kaptay, T. Matsushita, K. Mukai and T. Ohuchi, *Metallurgical and Materials Transactions B*, 35 (3) (2004), 471-486.

AUTHOR INDEX
Energy Technology Perspectives 2010

SUBJECT INDEX
Energy Technology Perspectives 2010

A

B

C

D

E

F

G

H

I

K

L

M

N

O

P

R

S

T

V

W

Z